Herausgeber:

Prof. Dr. *A. Davison*   Department of Chemistry, Massachusetts Institute
of Technology, Cambridge, MA 02139, USA

Prof. Dr. *M. J. S. Dewar*   Department of Chemistry, The University of Texas
Austin, TX 78712, USA

Prof. Dr. *K. Hafner*   Institut für Organische Chemie der TH
6100 Darmstadt, Schloßgartenstraße 2

Prof. Dr. *E. Heilbronner*   Physikalisch-Chemisches Institut der Universität
CH-4000 Basel, Klingelbergstraße 80

Prof. Dr. *U. Hofmann*   Institut für Anorganische Chemie der Universität
6900 Heidelberg 1, Tiergartenstraße

Prof. Dr. *K. Niedenzu*   University of Kentucky, College of Arts and Sciences
Department of Chemistry, Lexington, KY 40506, USA

Prof. Dr. *Kl. Schäfer*   Institut für Physikalische Chemie der Universität
6900 Heidelberg 1, Tiergartenstraße

Prof. Dr. *G. Wittig*   Institut für Organische Chemie der Universität
6900 Heidelberg 1, Tiergartenstraße

Schriftleitung:

Dipl.-Chem. *F. Boschke*   Springer-Verlag, 6900 Heidelberg 1, Postfach 1780

Springer-Verlag   6900 Heidelberg 1 · Postfach 1780
Telefon (06221) 49101 · Telex 04-61723
1000 Berlin 33 · Heidelberger Platz 3
Telefon (0311) 822001 · Telex 01-83319

Springer-Verlag   New York, NY 10010 · 175, Fifth Avenue
New York Inc.   Telefon 673-2660

Fortschritte der chemischen Forschung
Topics in Current Chemistry

Band 16, Heft 1, November 1970

Reactive Intermediates

Springer-Verlag
Berlin Heidelberg GmbH

ISBN 978-3-540-05103-9   ISBN 978-3-540-36331-6 (eBook)
DOI 10.1007/978-3-540-36331-6

Inhalt

Recent Aspects of the Chemistry of Sulphonyl Nitrenes

Prof. R. A. Abramovitch and **Prof. R. G. Sutherland**

Chemistry Department, University of Alabama, University, AL, USA,
and Chemistry Department, University of Saskatchewan, Saskatoon, Sask., Canada

Contents

1. Introduction

Nitrenes (imenes, azenes, imidogens, imido intermediates) are reactive intermediates having a monovalent nitrogen atom with a sextet of electrons in its outer shell. Sulphonyl nitrenes, RSO_2N, like other nitrenes and the isoelectronic carbenes can exist in either the singlet *(1)* or triplet state *(2)* and their chemical and physical properties

$$RSO_2\ddot{\underset{..}{N}} \; \rightleftharpoons \; RSO_2\dot{\underset{..}{N}}\cdot$$

1   *2*

will vary accordingly.

The chemistry of sulphonyl nitrenes has been briefly reviewed to the end of 1963 [1]. The present article concentrates mainly on selected subsequent advances in the generation and chemical behaviour of these reactive intermediates and on synthetic applications.

1

Whereas the intermediate produced by a thermal process has to be in the singlet state initially because of spin conservation, the ground state in this case appears to be the triplet *(2)*.

Thus, e. s. r. measurements have been reported [2,3] on the photolysis of benzene- and *p*-toluene-sulphonyl azide in a fluorolube matrix at 77 °K. No half-field transition was observed in either case, but broad (300 and 350 gauss) $\Delta M = 1$ transitions were observed at 7795 and 7740 gauss, respectively. These results were interpreted in terms of a triplet ground state or a relatively low-lying, thermally accessible, triplet level. Similar results were reported for the single crystal irradiation of *p*-fluorobenzenesulphonyl azide and methanesulphonyl azide at -160 °C with a medium pressure mercury arc [4]. In this case, hyperfine structure of the triplet *p*-fluorobenzenesulphonyl nitrene was observed: three equal lines with spacing of about 18 gauss centred at 8089 gauss at 9.2 kMc/sec. The zero-field splitting parameters were $D = 1.555$, $E < 0.005$ cm^{-1}. For triplet CH_3SO_2N, $D = 1.569$, $E = 0$ cm^{-1}. The resonance signals for *p*-fluorobenzenesulphonyl nitrene bleached at about -35 °C.

While the e. s. r. evidence points strongly to a triplet ground state, it tells us nothing about the state the sulphonyl nitrene is in when it undergoes reaction. Evidence bearing upon the latter comes from a consideration of the nature and relative rates of the reactions undergone by sulphonyl nitrenes and of the products obtained therefrom. This is considered in some detail in Section 3.3.

2. Generation of Sulphonyl Nitrenes

The production of the nitrenes from a variety of sources has been reported, the most common of which are the sulphonyl azides.

2.1. From Sulphonyl Azides

a) Thermolysis

Uncatalyzed thermal decomposition of sulphonyl azides is believed to give nitrene intermediates and nitrogen, in general:

$$RSO_2N_3 \xrightarrow{\Delta} RSO_2N + N_2$$

Until recently, most of the evidence for the rate-determining formation of a nitrene intermediate came from experiments in which the nitrene was trapped or from the temperatures required to effect decomposition and the nature of the products formed. Horner and Christmann [5] observed that the rate of nitrogen evolution from *p*-toluenesulphonyl azide at 130—135 °C was apparently independent of the nature of the solvent, and the decomposition of $C_6H_5SO_2N_3$ has been reported [6,7] to follow first-order kinetics with $k_1 = 1.48 \times 10^{-3}$ min.$^{-1}$ at 126.8 °C in chlorobenzene, nitrobenzene or p-xylene.

Recently, three groups have reported kinetic studies of the thermal decomposition of aryl and alkyl sulphonyl azides [8-10]. The decomposition of p-toluenesulphonyl azides is first-order [$\Delta H^{\neq} = 36.5$ kcal./mole, $\Delta S^{\neq} = 7$ e.u.] in a variety of solvents and the average half-life at 155 °C is 33 min. [8]. The thermolysis of benzenesulphonyl azide in boiling cyclohexanone is also first-order with $\Delta H^{\neq} = 33$ kcal./mole and $\Delta S^{\neq} = 5.2$ e.u. [9], and the rate constants for p-substituted derivatives were said to correlate well in a Hammett plot [11]. On the other hand, both Leffler and Tsuno [7] and Takemoto, Fujita, and Imoto [10] found a practically negligible substituent effect ($\varrho = -0.1$). In all cases, therefore, the rate-determining step is loss of nitrogen, to form the electron-deficient nitrene intermediate in which, in agreement with the e. s. r. data on the triplet ground state, there is little interaction between the electrons on the nitrogen atom and the aromatic nucleus.

The decomposition of a number of aliphatic mono- and di-sulphonyl azides in diphenyl ether was again first-order and these compounds were somewhat more stable than the aromatic sulphonyl azides [8]. When the thermolyses were effected in mineral oil a good first-order plot was not obtained [8,12] and it was found that the evolved gas contained sulphur dioxide. The solution still showed the presence of azide after ca. 60 half-lives. The amount of sulphur dioxide formed was essentially independent of the decomposition temperature. p-Toluenesulphonyl azide gave 1.3% of SO_2 [benzenesulphonyl azide in hot cyclohexanone gave 2.4% of SO_2 [9] and mesitylene-2-sulphonyl azide in n-dodecane at 150 °C gave a 22% yield of SO_2 [13]] whereas ca. 20% was formed from the alkylsulphonyl azides. That the reaction leading to SO_2 evolution with the alkyl compounds was mainly a radical process was shown by carrying out the reaction in the presence of inhibitors, e.g. hydroquinone, when the yield of SO_2 dropped to 3—5% and good first-order kinetics were again observed. Thus, in addition to the nitrene-forming reaction, another process is taking place which has been formulated by Breslow and his coworkers [8] as a radical chain decomposition (Scheme 1):

$$R \cdot + RSO_2N_3 \longrightarrow RSO_2 \cdot + R'N_3$$

$$RSO_2 \cdot \longrightarrow R \cdot + SO_2$$

$$R \cdot + R''-H \longrightarrow RH + R'' \cdot$$

Scheme 1

3

A careful study of this reaction showed it was *not first-order* and that it was completed by the time half of the sulphonyl azide had decomposed. The nature of the radical initiator has not been established. One possibility is that it is the triplet nitrene produced by singlet to triplet conversion from the initially formed singlet. This appears unlikely since the triplet might then be expected to be formed continuously during the run and not stop after an initial reaction period, unless either an inhibitor of the singlet → triplet conversion process were formed or a radical trap were generated in the reaction. The C—H insertion product formed could well act as such a radical trap. Leffler and Tsuno [6] reported the mutual decomposition of benzenesulphonyl azide and *t*-butyl hydroperoxide in chlorobenzene at 126.7 °C. Initially, the gas evolution is that expected of an uncatalyzed azide reaction, followed by a very fast reaction involving the evolution of nitrogen plus oxygen, followed by a slower uncatalyzed reaction. The induced decomposition may be prevented by the addition of iodine or changing the solvent to *p*-xylene. No sulphonamide is found using the latter solvent. These results suggest that the decomposition of azides to nitrenes can be induced by free radicals.

More recent work [8] shows that the S—N bond can be cleaved by hydroperoxides and that aromatic sulphonyl azides only undergo free radical thermal decomposition if a source of radicals is provided. Some light on the nature of the *radical transfer agent* has recently been shed by the observation [14] that dodecyl azides are formed (2.3%) in the thermolysis of mesitylene-2-sulphonyl azide *(3)* at 150 °C in *n*-dodecane under nitrogen. It seems likely that a dodecyl radical is produced by hydrogen abstraction by the triplet nitrene *(5)* [mesitylene-2-sulphonamide was also formed (1.1%)] which then attacks undecomposed sulphonyl azide as in Scheme 1, suggested by Breslow *et al.* [8]. Much of the SO_2 evolved, however, undoubtedly arises from an unstable sulphonyl aniline derivative 6 produced by a Curtius-type rearrangement; 6 then decomposes to SO_2 and the aryl nitrene 7. The latter undergoes hydrogen-abstraction or dimerization to give the aniline and the azobenzene derivative, respectively [14]. It is not known whether any Curtius-type rearrangement

products are formed in any of the decompositions of the aliphatic sulphonyl azides, and this should be looked into as a possible source of some of the sulphur dioxide formed in these reactions. Other studies on the radical-induced decomposition of sulphonyl azides have been reviewed briefly [15].

Evidence for the formation of alkyl and aryl radicals in some cases following loss of SO_2 (Scheme 1) has been obtained. Thus, a small amount of *n*-pentane was formed in the decomposition of *n*-pentanesulphonyl azide in mineral oil [8]. Thermolysis of diphenyl sulphone-2-sulphonyl azide (8) in dodecane at 150 °C gave diphenyl sulphone 9 (27%) and diphenyl sulphone-2-sulphonamide 10 (9%) which arise by hydrogen abstraction by the aryl radical and sulphonyl nitrene, respectively. When this thermolysis was carried out in Freon E-4 at 150 °C, the products were diphenylene sulphone 11 (1.3%) (Pshorr-type cyclization product of the aryl radical) and 10 (1.5%) together with tars [16]. Ferro-

5

cenylsulphonyl azide also gives some ferrocene on thermal decomposition in cyclohexane under pressure [17].

The rate of decomposition of benzenesulphonyl azide to benzenesulphonamide is said to be accelerated appreciably by thiophenol [18] in a radical-catalyzed process probably not involving a free nitrene intermediate.

b) Photolysis

Though low-temperature photolysis of sulphonyl azides in a frozen matrix [2,3] or as single crystals [4] has been shown to result in the production of a triplet nitrene, irradiation of sulphonyl azides in solution in nonprotic non-polar solvents usually gives rise to tar formation with little or no evidence for the intervention of nitrenes. Thus, photolysis of sulphonyl azides in benzene, cyclohexene, pyridine and thiophene generally gives rise to insoluble polymer-like products [5,19,20]. For example, photolysis of methanesulphonyl azide in benzene using a medium pressure arc, gave a yellow, amorphous material which did not melt below 290 °C, was sparingly soluble in ethanol and formed a gum with boiling ethanol. On the other hand, if care is taken to prevent the deposition of polymer on the sides of the quartz flask, photolysis at either 28 °C or 80 °C gave a small amount of nitrene which reacted with the benzene solvent [19] (see Section 3.3). Photolysis of benzenesulphonyl azide in benzene gave mainly undecomposed azide, a dark amorphous material and some benzenesulphonylaniline and benzenesulphonamide identified by infrared spectroscopy [20].

Photolysis of sulphonyl azides in dimethyl sulphoxide with 2537 Å light gives N-sulphonylsulphoximines 12 in 15—50% yield [5]. The reaction was formulated as going *via* a nitrene intermediate which was trapped by the nucleophilic solvent

$$RSO_2N_3 \xrightarrow{h\nu} N_2 + RSO_2\ddot{N} \xrightarrow{(CH_3)_2SO} RSO_2N=\overset{O}{\overset{\uparrow}{S}}(CH_3)_2$$
$$12$$

Dimethyl sulphide was a better trapping agent than dimethyl sulphoxide and 48—55% yields on N-sulphonylsulphimines 13 were obtained [5].

$$RSO_2N_3 + Me_2S \xrightarrow{h\nu} RSO_2N=SMe_2 + N_2$$
$$13$$

Though the intermediacy of nitrenes is likely in these reactions, the possibility of prior complexing of the sulphonyl azide with the solvent acting as an electrophile *followed* by nitrogen elimination exists with dimethyl sulphoxide and has not been disposed of. This mechanism cannot obtain with dimethyl sulphide, however, and a discrete nitrene intermediate mechanism seems the most probable in this case.

$$RSO_2-\overset{-}{N}-\overset{+}{N}\equiv N + Me_2\overset{+}{S}-\overset{-}{O} \longrightarrow \left[RSO_2-N-\overset{+}{N}\equiv N \atop \overset{-}{\underset{O-SMe_2}{C}} \right] \text{ or } \left[RSO_2-N \overset{N}{\underset{Me_2S\underline{\quad\quad}O}{\diagdown}} N \right]$$

$$\xrightarrow{h\nu} RSO_2N=\overset{O}{\underset{}{\overset{\uparrow}{S}Me_2}} + N_2$$

In contrast to the behaviour in non-polar solvents, photolysis of sulphonyl azides *in alcohols* occurs readily [5,20,21] to give products of insertion into the alcohols, hydrogen abstraction and Curtius-type rearrangement. Hydrogen bonding by the solvent appears to be important. Thus, irradiation of benzenesulphonyl azide in methanol may not involve a free nitrene intermediate. The latter could be formed and rapidly protonated before undergoing rearrangement; alternatively, a concerted rearrangement of a hydrogen-bonded sulphonyl azide is possible, so that a free nitrene is never formed [20].

$$C_6H_5\overset{O}{\underset{O}{\overset{\parallel}{\underset{\parallel}{S}}}}-\overset{-}{N}-\overset{+}{N}\equiv N \xrightarrow{CH_3OH} \underset{Ph}{\overset{\delta^-..H...OCH_3}{\underset{O}{\overset{\delta^-}{\underset{O}{\diagup}}}}S\overset{\delta^+}{\underset{\delta^+}{\cdots}}N\cdots N_2} \xrightarrow{-N_2} O=S=NPh \overset{..HOCH_3}{\underset{}{\overset{O^-}{}}}$$

The reaction of methanesulphonyl azide in isopropanol, either with 2537 Å light or using 3660 Å light and a benzophenone sensitizer, proceeds differently, and leads to a quantitative yield of methanesulphonamide and acetone [21]. The direct photolysis of methanesulphonyl azide gave quantum yields varying from 20 to 75 on runs taken to 20% completion. In the sensitized photolysis, low initial azide concentrations gave good first-order kinetics which held up to five-fold increases in concentration, but a ten-fold increase showed unmistakable departure from first-order behaviour. The quantum yield was much greater than unity and appeared to be an inverse function of light intensity, the rate law containing both half-order and first-order light intensity terms. A *radical chain mechanism* consisting of two propagation sequences was proposed, with the possible exclusion of Eq. (2) or Eqs. (5) and (6).

$$Ph_2CO \xrightarrow{h\nu} Ph_2CO^* \tag{1}$$

$$Ph_2CO^* + Me_2CHOH \longrightarrow Ph_2\dot{C}OH + Me_2\dot{C}OH \tag{2}$$

$$Ph_2CO + Me_2\dot{C}OH \longrightarrow Ph_2\dot{C}OH + Me_2CO \tag{3}$$

$$2 Ph_2\dot{C}OH \longrightarrow \underset{\underset{OH}{|}}{Ph_2C} - \underset{\underset{OH}{|}}{CPh_2} \tag{4}$$

$$Ph_2CO^* + MeSO_2N_3 \longrightarrow Ph_2CO + MeSO_2\ddot{\underset{..}{N}}\cdot + N_2 \tag{5}$$

$$MeSO_2\ddot{\underset{..}{N}}\cdot + Me_2CHOH \longrightarrow MeSO_2\dot{\underset{..}{N}}H + Me_2\dot{C}OH \tag{6}$$

$$MeSO_2\dot{\underset{..}{N}}H + Me_2CHOH \longrightarrow MeSO_2NH_2 + Me_2\dot{C}OH \tag{7}$$

$$MeSO_2N_3 + Me_2\dot{C}OH \longrightarrow MeSO_2\dot{\underset{..}{N}}H + Me_2CO + N_2 \tag{8}$$

$$MeSO_2N_3 + Ph_2\dot{C}OH \longrightarrow MeSO_2\dot{\underset{..}{N}}H + Ph_2CO + N_2 \tag{9}$$

Interestingly, no SO_2 was evolved in this reaction as in the photolysis of α-toluenesulphonyl azide. This could be explained on the basis of a cation-radical anion pair which collapses as in Eq. (8) to give a sulphonamido radical, and no free nitrene is formed [21].

$$MeSO_2N_3 + R_2\dot{C}OH \longrightarrow [MeSO_2\ddot{\underset{..}{N}}\ \ddot{N}=\overset{+}{N}\cdot R_2\dot{C}OH] \xrightarrow{-N_2} MeSO_2\dot{\underset{..}{N}}H + R_2CO$$

The decomposition of methanesulphonyl azide in isopropyl alcohol could be effected by selective irradiation of 2-acetonaphthone instead of benzophenone [21]. Since 2-acetonaphthone triplets are incapable of hydrogen abstraction from isopropyl alcohol [22], initiation must occur *via* transfer of excitation energy to the azide. A marked difference was observed from benzophenone sensitization in that the reaction was extremely slow, gave a nitrogen yield of only 68%, and produced a yellow solution [21].

Ferrous chloride-hydrochloric acid mixtures catalyzed the thermal decomposition of sulphonyl azides in isopropyl alcohol to give occasionally almost quantitative yields of sulphonamide and acetone, and the molar ratio of azide consumed to ferric chloride formed was typically of the order of 20 to 1 [21].

$$RSO_2N_3 + FeCl_2 + HCl \longrightarrow RSO_2\dot{\underset{..}{N}}H + FeCl_3 + N_2$$

The photolysis of benzenesulphonyl azide in thiophenol has also been reported to involve a free radical decomposition [18].

Horner and Bauer [23] initiated the thermal free radical decomposition of p-toluenesulphonyl azide in isopropanol with diethyl peroxydicarbonate and formulated the mechanism as follows:

$$(EtO\overset{\overset{O}{\|}}{C}-O)_2 \longrightarrow 2\ EtO\overset{\overset{O}{\|}}{C}O\cdot \longrightarrow 2\ EtO\cdot + CO_2$$

$$EtO\cdot + Me_2CHOH \longrightarrow EtOH + Me_2\dot{C}OH$$

$$C_7H_7SO_2N_3 + Me_2\dot{C}OH \longrightarrow C_7H_7SO_2N{=}N{-}\underset{\underset{OH}{|}}{\overset{\cdot}{N}}{-}CMe_2$$

$$C_7H_7SO_2\dot{N}H + N_2 + Me_2CO \longleftarrow C_7H_7SO_2\overset{\cdot}{N}\diagup\!\!\!{\overset{N}{\diagdown}}\!\!\!\diagdown N$$

$$C_7H_7SO_2\dot{N}H + Me_2CHOH \longrightarrow C_7H_7SO_2NH_2 + Me_2\dot{C}OH$$

In contrast to the problems encountered on photolysis of alkyl- and aryl-sulphonyl azides, we have found that ferrocenylsulphonyl azide *14* is smoothly decomposed by 3500 Å light in cyclohexane or in benzene to give ferrocene *15*, ferrocenylsulphonamide *16* and the novel bridged [2]ferrocenophanethiazine 1,1-dioxide *17* [24]. The yield of *17* varied with the nature of the solvent, being 13.3% in cyclohexane, 67% in benzene, and zero in dimethyl sulphoxide or DMSO/benzene [25].

The bridged compound is not formed at all on thermolysis of *14*: instead, the main products are *15*, *16* and the substituted amide [17].

$$FcSO_2N_3 \xrightarrow[\text{RH}]{\Delta} 15 + 16 + FcSO_2NHR$$

14

RH			
$RH = C_6H_{12}$	20%	48.4%	24%
$RH = C_6H_6$	17%	76.8%	6.5%
$RH = C_6H_{10}$	2%	85.4%	8%

9

As will be discussed later, it is possible [24] that the thermolysis involves a metal-nitrene complex whereas the photolysis involves the free nitrene. The product distribution is not affected by the presence of a photosensitizer, but since ferrocene itself is both an efficient triplet quencher as well as a sensitizer [26,27] it is very difficult to probe the spin state of ferrocenyl nitrene at the moment of reaction. The cyclization appears to be a singlet reaction since the yield of 27 in benzene solution is essentially unaffected by oxygen or the presence of hydroquinone [25].

c) Metal-catalyzed Decompositions

Despite the volume of work concerned with metal-catalyzed decomposition of diazo compounds and "carbenoid" reactions [28], relatively little work has been reported on the metal-catalyzed decomposition of sulphonyl azides. Some metal-aryl nitrene complexes have recently been isolated [29-31]. Nitro compounds have also been reduced to nitrene metal complexes with transition metal oxalates [32].

Kwart and Khan investigated the copper-catalyzed decomposition of benzenesulphonyl azide both in methanol [33] and in cyclohexene [34]. No reaction occurs between benzenesulphonyl azide and cyclohexene at 100 °C but the addition of copper powder causes a smooth decomposition to take place yielding an impressive array of products [34]. The major ones are benzenesulphonamide 18 (37%), the aziridine 19 (15%) and the N-(1-cyclohexenyl)benzenesulphonamide 20 (17%) (Scheme 2). Some traces of cyclohexyl azide were also found but the addition of hydroquinone eliminated its formation.

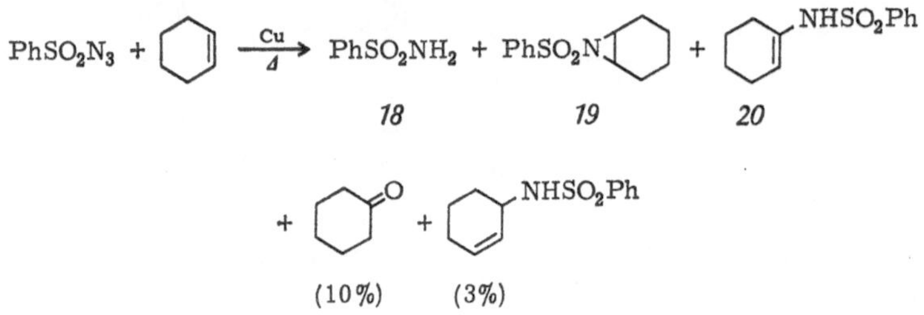

Scheme 2

The product distribution was dramatically affected by the addition of dimethyl sulphoxide to the reaction mixture:

$$PhSO_2N_3 + \text{[benzene ring]} \xrightarrow[DMSO]{Cu} \underset{25\%}{\overset{18}{}} + \underset{13\%}{\overset{20}{}} + PhSO_2N=\overset{O}{\underset{21}{\underset{50\%}{S}Me_2}} + \underset{15\%}{\overset{O}{\text{[cyclohexanone]}}}$$

The addition of copper to a boiling solution of benzenesulphonyl azide in methanol gave benzenesulphonamide *18* (80%) as the major product together with minor amounts of methylenebis(benzenesulphonamide) *22* and 1,3,5-tris(benzenesulphonyl)hexahydro-s-triazine *23*, resulting from condensation of *18* with the formaldehyde formed in the reaction [33]. Cuprous chloride was even more effective, but cuprous

$$PhSO_2N_3 + CH_3OH \xrightarrow[\Delta]{Cu} PhSO_2NH_2 + [HCHO] + N_2$$

$$PhSO_2NH_2 + HCHO \longrightarrow (PhSO_2NH)_2CH_2 + PhSO_2N \underset{23}{\overset{\overset{N-SO_2Ph}{\diagup}}{\underset{N-SO_2Ph}{\diagdown}}}$$

oxide was inert. The decomposition also occurred in isopropanol to give an almost quantitative yield of *18* and some acetone, and in wet t-butanol. Insoluble green copper complexes were always formed. The authors proposed that a copper azide complex *(24)* is involved which loses nitrogen to give a transient copper nitrene complex *(25)*.

$$\underset{\underset{\underset{24}{SO_2Ph}}{\overset{|}{}}}{Cu \overset{\ddot{N}}{\underset{\ddot{N}}{\diagup \diagdown}} N:} \xrightarrow{-N_2} [Cu \overset{\pm}{=} N-SO_2Ph \rightleftharpoons \underset{25}{Cu^\circ \ldots \cdot NSO_2Ar}]$$

Dimethyl sulphoxide (amounting to slightly more than equimolar with azide and less than 1% overall concentration in solution in methanol) accelerates the copper-catalyzed decomposition and the only product formed (97%) is the sulphoximine *(21)*. Even in the absence of copper, DMSO and benzenesulphonyl azide were found to undergo a slow reaction in boiling methanol to give *21* (< 40%). It was suggested [33] that the sulphonyl azide itself (slow) or the copper complex *24* (fast) underwent 1,3-dipolar addition with DMSO to give an oxathiatriazoline which lost nitrogen to give *21*, so that no free nitrene was formed in this reaction.

Copper catalyzes the decomposition of sulphonyl azides in benzene very slowly. When methanesulphonyl azide was boiled under reflux in benzene solution in the presence of an excess of freshly reduced copper powder, some decomposition occurred to give methanesulphonamide and azide was recovered [78]. Transition metal complexes have been found to exert a marked effect upon the yields of products and isomer ratios formed in the thermal decomposition of methanesulphonyl azide in methyl benzoate and in benzotrifluoride [36]. These results will be discussed in detail in the section on the properties of sulphonyl nitrenes and singlet and triplet behaviour. A sulphonyl nitrene-iron complex has recently been isolated [37] and more on this species will be reported soon.

2.2. Other Sources of Sulphonyl Nitrenes

Starting materials other than sulphonyl azides have been used as possible sources of sulphonyl nitrenes. The decomposition of the triethylammonium salt of N-p-nitrobenzenesulphonoxybenzenesulphonamide *(26)* in methanol, ethanol, and aniline gave products derived from a Lossen-type rearrangement [20] (Scheme 3). It was felt that the rearrangement did not involve a free sulphonyl nitrene since, when the decomposition was carried out in toluene-methylene chloride or in benzene, no products (benzenesulphonamides) of substitution of the aromatic solvent nucleus were found (as are usually found with sulphonyl nitrenes from the thermal decomposition of the corresponding azides). On the other

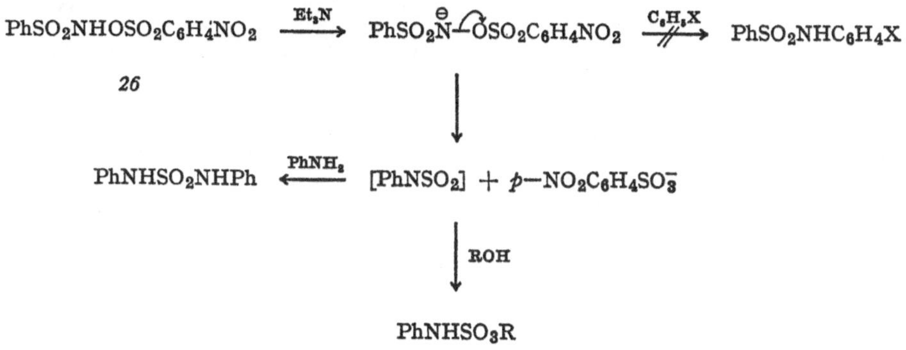

$$PhSO_2NHOSO_2C_6H_4NO_2 \xrightarrow{Et_3N} PhSO_2\overset{\ominus}{N}\overset{\frown}{-}OSO_2C_6H_4NO_2 \xrightarrow{C_6H_5X} PhSO_2NHC_6H_4X$$

26

$$\downarrow$$

$$PhNHSO_2NHPh \xleftarrow{PhNH_2} [PhNSO_2] + p{-}NO_2C_6H_4SO_3^-$$

$$\downarrow ROH$$

$$PhNHSO_3R$$

Scheme 3

hand, since the reactions were carried out at room temperature, the products expected [38] would be *N*-sulphonylazepines and not the anilides, so that this reaction is worthy of re-investigation. In fact, the base-initiated decomposition of *N*-arylsulphonoxysulphonamides has recently been looked into again and it has been shown that a nitrene intermediate is probably involved that can be trapped by dimethylsulphide and, less efficiently, by dimethyl sulphoxide [39].

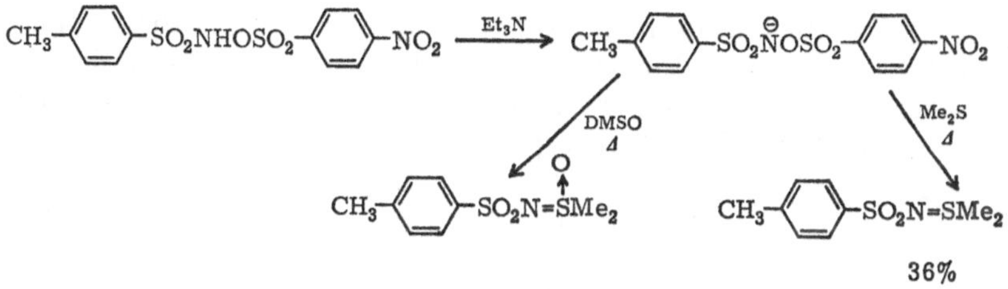

The thermal decomposition of *N*-trimethylammoniododecanesulphon-amide *(27)* in dimethyl sulphoxide at 170 °C gives dodecanesulphon-amide *(28)* and the sulphoximine *29*, as would be expected from a sulphonyl nitrene intermediate [40]. Photolysis of *27* did not take place, even at 80 °C though some *29* was obtained on irradiation in DMSO. It is not clear whether or not *27* actually absorbs in the ultraviolet region of interest. Sulphonamidate *27* decomposes in decalin at 175 °C in the presence of triphenylphosphine to give *28* and the phosphinimide *30*. The latter was the sole product when a five-fold excess of Ph₃P was used. No C—H insertion products were found, which is rather surprising in view of the results of Breslow *et al.* on the behaviour of alkylsulphonyl nitrenes in hydrocarbon solvents [8,12].

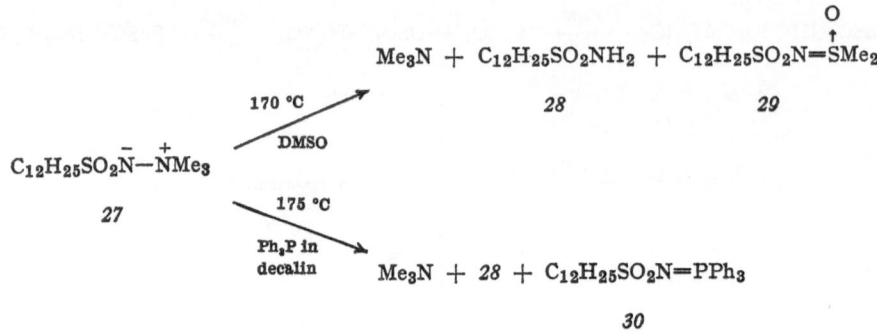

On the other hand, irradiation of the pyridinium ylid *31* using a Pyrex filter [41] gave the corresponding 1-arylsulphonyl-1,2-diazepine *32*. No evidence for the formation of any nitrene intermediates in this reaction was reported. It has been confirmed that photolysis of *31* (Ar = Ph) with 3000 Å light in benzene-acetonitrile gave a 76% yield of *32* (Ar = Ph) and none of the products expected of a free nitrene in solution: benzene-sulphonamide or *N*-benzenesulphonylazepine [42]. On the other hand, if the photolysis was carried out in DMSO solution, the yield of *32* dropped to 12% and a 40% yield of the sulphoximine *27* was isolated. Photolysis of *31* (Ar = *p*-CH$_3$C$_6$H$_4$) in benzene-acetonitrile solution using 3500 Å light gave low yields of the diazepine *32* (Ar = *p*-CH$_3$C$_6$H$_4$) and of *p*-toluenesulphonamide in the ratio of about 2:1 [42]. Thus, it would appear that under certain conditions sulphonyl nitrenes may indeed be formed by photolysis of suitable pyridinium ylids (the reaction in DMSO may involve an intermediate of the type

$$ArSO_2-N-\overset{+}{N}C_5H_5$$
$$|$$
$$^{-}O-SMe_2$$

and a free nitrene may not be generated) and further work is being carried out along those lines.

$$ArSO_2\overset{-}{N}-\overset{+}{N}\hspace{1em}\xrightarrow{h\nu}\hspace{1em}ArSO_2N$$

31   *32*

Photolysis (2537 Å) of *N*-arylsulphonyldimethyl sulphoximines in aromatic hydrocarbon solvents did not produce arylsulphonyl nitrenes: instead, aryl radicals were generated which arylated the solvent [43].

$$ArSO_2N \xleftarrow{\;h\nu\;}\!\!\!\!/\!\!\!\!/ \;\; ArSO_2 \overset{\overset{O}{\uparrow}}{N}{=}SMe_2 \xrightarrow[C_6H_6]{h\nu} [Ar\cdot] \longrightarrow ArC_6H_5$$

Recently, Breslow and Sloan [44] reported that the reaction of dichlor-amine-T *(33)* with zinc dust in cyclohexane gave the insertion product, *N*-cyclohexyl-*p*-toluenesulphonamide *(34)* in 80% yield. This was for-mulated as follows:

$$p\text{-MeC}_6\text{H}_4\text{SO}_2\text{NCl}_2 \xrightarrow{\;Zn\;} p\text{-MeC}_6\text{H}_4\text{SO}_2\overset{\overset{\text{Cl}}{|}}{\text{N}}{-}\text{ZnCl} \xrightarrow{-ZnCl_2}$$

33

$$p\text{-MeC}_6\text{H}_4\text{SO}_2\overset{..}{\underset{..}{\text{N}}} \xrightarrow{C_6H_{12}} p\text{-MeC}_6\text{H}_4\text{SO}_2\text{NHC}_6\text{H}_{11}$$

34

The success of this reaction was ascribed to the solubility of the chlorozinc intermediate, whereas other chloramine-T derivatives (e. g. the sodium salt) are insoluble. An alternative non-nitrene pathway was not eliminated from consideration. On the other hand, no aromatic substitu-tion or addition, characteristic of a free sulphonyl nitrene (see below), took place on treatment of *N,N*-dichloromethanesulphonamides with zinc powder in benzene in the cold or on heating. The only product isolated was that of hydrogen-abstraction, methanesulphonamide [42], which appears to be more characteristic of the behaviour of a sulphonyl nitrene-metal complex [36,37]. Photolysis of *N,N*-dichloromethanesulphonamide, or dichloramine-B, or dichloramine-T in benzene solution led to the formation of some unsubstituted sulphonamide and some chlorobenzene but no product of addition of a nitrene to benzene [19].

A copper-sulphonyl nitrene complex has been postulated as an intermediate in the reaction of chloramine-T with DMSO and with dioxan [45]. In the absence of copper powder, only a small yield of sulphoximine (6%) was obtained in DMSO. Again, no sulphonylaniline or azepine

$$p\text{-MeC}_6\text{H}_4\text{SO}_2 \overset{\overset{O}{\uparrow}}{N}{=}SMe_2 \xleftarrow[Cu]{DMSO} p\text{-MeC}_6\text{H}_4\text{SO}_2\overset{-}{N}\text{Cl}\overset{+}{N}a \xrightarrow[Cu]{}$$

$$p\text{-MeC}_6\text{H}_4\text{SO}_2\text{NH}{-}$$

derivative was isolated from the reaction of chloramine-T with copper powder in benzene. Only the unsubstituted sulphonamide was obtained [42].

In contrast to the reaction of *benzamide* and other carboxylic acid amides with *lead tetraacetate*, which has been said to proceed *via* an acyl nitrene intermediate [46], methanesulphonamide and 2-biphenylsulphonamide are completely inert towards this reagent [42].

Finally, a sulphonyl nitrene has been suggested as a possible intermediate in the reaction of *benzenesulphonyl isocyanate* with *phenylmagnesium bromide* at room temperature, to account for the formation of a 13% yield of *N*-phenylbenzenesulphonamide [47].

$$PhSO_2N{=}C{=}O \longrightarrow CO + PhSO_2N \xrightarrow{\text{PhMgBr}} PhSO_2\overset{\overset{\displaystyle Ph}{|}}{N}MgBr \xrightarrow{\text{H}_2\text{O}}$$

$PhSO_2NHPh$

This seems highly unlikely in view of the very mild reaction conditions. On the other hand, the photolysis of the isocyanate may yield the desired nitrene.

3. Reactions

As with carbenes and with other nitrenes, the fate of the reactive intermediate RSO_2N depends on the solvent, the nature of R, the spin state of the nitrene at the moment of reaction, and the presence or absence of metals. Though the intermediate is usually generated as the singlet, since the ground state is the triplet, intersystem crossing can occur if the intermediate survives several collisions or if the substrate is relatively inert. Thus, to determine the spin state at the moment of reaction requires these effects to be unravelled. It is also important to bear in mind that the reaction observed may not be due to a nitrene intermediate but to a precursor such as the sulphonyl azide, e. g. triazoline formation followed by loss of nitrogen to give the same product as would be derived from the free nitrene.

The chemical reactions of sulphonyl nitrenes include hydrogen abstraction, insertion into aliphatic C—H bonds, aromatic "substitution", addition to olefinic double bonds, trapping reactions with suitable nucleophiles, and Wolff-type rearrangement. Hydrogen-abstraction from saturated carbon atoms is usually considered to be a reaction typical of triplet nitrenes [8,48)], while stereospecific C—H insertion is thought to be typical of singlet nitrenes [8,49)]. Singlet carbethoxynitrene expands a benzene ring to the corresponding azepine, while triplet does not [50)].

3.1. Hydrogen Abstraction

Alkylsulphonyl nitrenes undergo hydrogen-abstraction to a small extent when generated in the presence of aliphatic hydrocarbons. Thus, 2-propanesulphonyl azide gave a maximum of 3% of 2-propanesulphonami-de on thermolysis in cyclohexane, while hydrogen-abstraction by 1-pentanesulphonyl nitrene was not proved [8)]. α-Toluenesulphonyl nitrene gave a 26.5% yield of α-toluenesulphonamide by abstraction from *n*-dodecane [16)], 1-pentylsulphonyl nitrene gave 1% of dicumyl by abstraction from the side-chain of cumene [8)], and methanesulphonyl nitrene gave 0.5% of dibenzyl by abstraction from the side-chain of toluene [51)]. *n*-Dodecylsulphonyl nitrene abstracted hydrogen at 170 °C in DMSO [40)].

Arylsulphonyl nitrenes usually give better yields of hydrogen-abstraction products from aliphatic hydrocarbons. *p*-Toluenesulphonyl azide gave a 5% yield of *p*-toluenesulphonamide on thermolysis in cyclohexane [8)] [a 40% yield was reported, however, by Pritzkow and Timm [52)]]. 2-Phenoxybenzenesulphonyl azide in dodecane at 135 °C gave phenoxy-benzenesulphonamide (4%) among other products; diphenyl sulphide 2-sulphonyl azide gave a 19% yield of the hydrogen-abstraction product, diphenyl sulphone-2-sulphonyl azide a 9% yield [16)], while mesitylene-2-sulphonyl azide in dodecane at 150 °C gave a 1.1% yield of sulphonamide [14)].

On the other hand, thermolysis of *ferrocenylsulphonyl azide (14)* in aliphatic solvents may lead to the predominant formation of the amide *(16)* [17)]. A 48.4% yield of *(16)* was obtained from the thermolysis in cyclohexane while an 85.45% yield of *16* was formed in cyclohexene. Photolysis of *14* in these solvents led to *lower* yields of sulphonamide: 32.2% in cyclohexane, 28.2% in cyclohexene. This suggests again that a metal-nitrene complex is an intermediate in the thermolysis of *14* since hydrogen-abstraction appears to be an important mode of reaction for such sulphonyl nitrene-metal complexes. Thus, benzenesulphonamide was the main product (37%) in the copper-catalyzed decomposition of the azide in cyclohexane, and the yield was not decreased (in fact, it increased to 49%) in the presence of hydroquinone [34)]. On the other hand, no toluene-sulphonamide was reported from the reaction of dichloramine-T and zinc in cyclohexane.

Hydrogen-abstraction from *aromatic solvents* occurs more readily and has been reported often [51,53—55)]. For example, the thermolysis of ben-zenesulphonyl azide in benzene at 105—120 °C gave an 18% yield of benzenesulphonamide [55]) while methanesulphonyl azide in benzene at 120 °C gave methanesulphonamide (14.4%) [51]). This is not a simple radical abstraction reaction by the triplet nitrene. Thus, only 0.4% yield of dibenzyl was formed in the reaction of $CH_3SO_2N_3$ with toluene, much less than the 22—24% of $CH_3SO_2NH_2$ also isolated. The methanesulphonamide does not result to any extent from hydrogen-abstraction by the nitrene from the sulphonylanilide formed since, in the reaction of methanesulphonyl azide with an excess of toluene at 120 °C, the yield of methanesulphonamide was 22.7% and that of the three N-mesyltoluidines 76.8%, thus accounting for 99.5% of the methanesulphonyl azide used [51]). Hydrogen-abstraction must be taking place from the aromatic nucleus as the only other source of hydrogen. That the nitrene is not abstracting one hydrogen atom at a time is shown by the fact that in none of the examples were any biphenyl derivatives detected [51]), which would have been the case had abstraction of a single hydrogen atom to give an aryl radical taken place (Scheme 4).

$$CH_3SO_2\overset{\cdot\cdot}{\underset{\cdot\cdot}{N}}\cdot + C_6H_6 \longrightarrow CH_3SO_2\overset{\cdot}{N}H + C_6H_5\cdot$$

$$C_6H_5\cdot + C_6H_6 \longrightarrow [C_6H_5C_6H_6\cdot] \xrightarrow{CH_3SO_2\overset{\cdot}{N}H} C_6H_5C_6H_5 + CH_3SO_2NH_2$$

Scheme 4

Consequently, abstraction of both hydrogens must be taking place in a *concerted*, or an almost concerted, process, and it has been suggested that the initially formed aziridine intermediate may collapse to the sulphonamide and a benzyne, the latter going to tars [51]). No evidence has been presented or obtained for the intervention of benzynes in these reactions. The simultaneous abstraction of two neighboring hydrogen atoms has

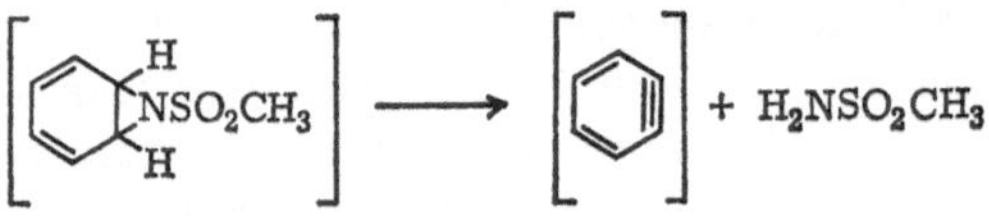

$$\left[\underset{H}{\overset{H}{\underset{\displaystyle }{\bigcirc}}}NSO_2CH_3\right] \longrightarrow \left[\bigcirc\right] + H_2NSO_2CH_3$$

been suggested in the carbene field [56]) and may

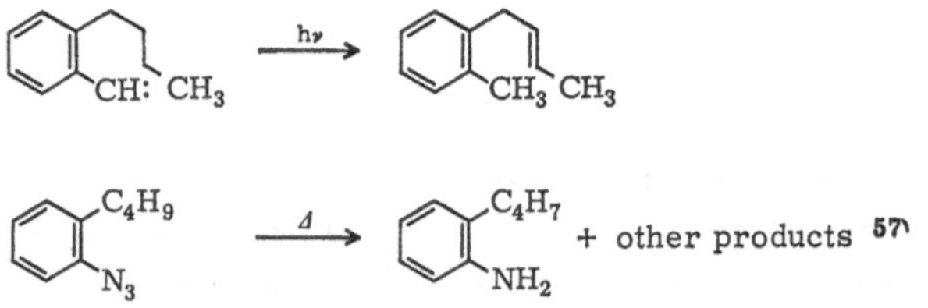

18

well be important in the reactions of other nitrenes (e. g. carbethoxy- and aryl-nitrenes).

Sulphonyl nitrene-metal complexes also abstract hydrogen from aromatic solvents in what may be an important reaction in these solvents. For example, whereas the decomposition of methanesulphonyl azide in degassed benzotrifluoride under nitrogen at 120 °C gives $MeSO_2NHC_6H_4CF_3$ (20.4%) and $MeSO_2NH_2$ (21.9%). The addition of $Fe_3(CO)_{12}$ to the original reaction mixture leads to a drop in anilide yield to 0.75% and an increase in methanesulphonamide yield to 61.5% [36]. Thermolysis of ferrocenylsulphonyl azide in benzene gave the sulphonamide (76.8%) [17], and methanesulphonamide is formed in the reaction of $MeSO_2NCl_2$ with zinc in benzene [42].

3.2. Insertion into Aliphatic C—H Bonds

Examples of such reactions are well known. Sloan, Breslow, and Renfrow found that both alkane and arenesulphonyl azides insert into the carbon-hydrogen bonds of saturated hydrocarbons [12]. Thus, 1-pentane,- 2-propane- and *p*-toluene-sulphonyl nitrene inserted into cyclohexane to give 54, 60, and 58% yields of the corresponding *N*-cyclohexylamide derivatives [8]. Similarly, 2-phenoxybenzene-, diphenyl sulphide-2-, and

$$RSO_2N_3 + C_6H_{12} \xrightarrow{\Delta} RSO_2NHC_6H_{11} + N_2$$

α-toluene-sulphonyl azides gave, on thermolysis in *n*-dodecane, the corresponding *N*-dodecyl insertion products [16], as did mesitylene-2-sulphonyl azide [14]. Thermolysis of ferrocenylsulphonyl azide in cyclohexane gave *N*-cyclohexylferrocenyl-sulphonamide (24%) [17]. Two examples of intramolecular insertion into aliphatic C—H bonds have been reported. Mesitylene-2-sulphonyl azide *(3)* [14] and durene-3-sulphonyl azide *(35)* [16] gave the corresponding sultams 36 and 37 in 2 and 15% yield, respectively, on thermolysis in dodecane at 150 °C.

36

35 *37*

Only limited success was achieved in determining the relative reactivity of primary, secondary, and tertiary carbon-hydrogen bonds to sulphonyl nitrenes [8]. Insertion of p-toluenesulphonyl nitrene into 2-methylbutane gave a mixture of products which could not be completely resolved. The ratio of (primary):(secondary + tertiary) = [38 + 39:40 + 41] was 1.53, compared to a ratio of 5.6 for carbethoxynitrene [58], indicating the lowered selectivity of the sulphonyl nitrene relative to the carbethoxynitrene, as might be expected from the possible resonance stabilization of the latter species.

$$RSO_2N_3 + (CH_3)_2CHCH_2CH_3 \longrightarrow (CH_3)_2CHCH_2CH_2NHSO_2R$$

38

$$+ RSO_2NHCH_2-\overset{\overset{\displaystyle CH_3}{|}}{C}HCH_2CH_3 + (CH_3)_2CHCH-CH_3$$

$$\underset{\displaystyle |}{NHSO_2R}$$

39   40

$$+ (CH_3)_2CCH_2CH_3$$

$$\underset{\displaystyle |}{NHSO_2R}$$

41

No insertion product was observed on photolysis of ferrocenylsulphonyl azide in cyclohexane or in cyclohexene [25], suggesting that the reactive intermediate formed is the triplet sulphonyl nitrene. The fact that addition to the olefinic bond of cyclohexene takes place under these conditions [25] does not necessarily argue against this conclusion (*vide infra*).

Sulphonyl nitrene-metal complexes also undergo insertion into aliphatic C—H bonds as witnessed by the insertion into dioxan on treatment with chloramine-T and copper [45] and into cyclohexane with dichloramine-T and zinc [44] and into cyclohexene with benzenesulphonyl azide and copper [34] and with ferrocenylsulphonyl azide [25].

3.3. Aromatic "Substitution"

This is the best known of the reactions of sulphonyl nitrenes and a large number of examples were studied by Curtius and his coworkers. The first attempt to establish semi-quantitative correlations between the nature of the substituent and orientation of the entering group was made by

Dermer and Edmison [54], and a somewhat more refined approach was taken by Heacock and Edmison [55] whose results on the benzenesulphonamidation of aromatic substrates are given in Table 1.

$$C_6H_5SO_2N_3 + C_6H_5X \xrightarrow{\Delta} C_6H_5SO_2NHC_6H_4X + C_6H_5SO_2NH_2 + N_2$$

Table 1. *Benzenesulphonamidation of aromatic substrates C_6H_5X at 105—120 °C* [55]

Substrate	Isomer Ratio			Total Rate Ratio	„Partial Rate Factors"		
	o	m	p	$\frac{X}{H}K$	F_o	F_m	F_p
$C_6H_5CH_3$	61	1	38	1.00	1.8	0.03	2.3
C_6H_5Cl	46	2	52	0.69	0.95	0.04	2.2
C_6H_5Br	50	5	45	0.69	1.0	0.10	1.9
$C_6H_5OCH_3$	71	2	27	0.96	2.0	0.06	1.6
$C_6H_5NH_2$	40	7	53	—	—	—	—
$C_6H_5CO_2CH_3$	43	54	3	0.38	0.49	0.62	0.07
C_6H_5OH	50	2	48	0.80	1.2	0.05	2.3
C_6H_5COCl	0	100	0	—	—	—	—

The above results were said to be consistent with an attack of the aromatic nucleus by the electrophilic free radical

$$C_6H_5SO_2\overset{\cdot}{\underset{\cdot\cdot}{N}}\cdot.$$

Abramovitch, Roy, and Uma [51] disagreed with this, pointing out a number of inconsistencies with that conclusion. Thus, while the total rate ratios are not much different from unity, as expected for a homolytic substitution, the values of $^{CH_3}_H K = 1.0$, $^{OMe}_H K = 0.96$, and $^{OH}_H K = 0.80$ do not support this mechanism since such electron-donating substituents should facilitate attack by an electrophilic free radical [59,60] and lead to total rate ratios greater than unity. Also, the partial rate factor calculated for attack at the *meta* position of toluene was unusually low, and it is not clear why this position should be deactivated towards attack either by a free radical or by an electrophilic species.

The thermolysis of methanesulphonyl azide in aromatic solvents at 120 °C was studied quantitatively and the results are summarized in Table 2.

$$CH_3SO_2N_3 + C_6H_5X \xrightarrow{\Delta} CH_3SO_2NHC_6H_4X + CH_3SO_2NH_2 + N_2$$

21

Table 2. *Methanesulphonamidation of C_6H_5X at 120° under nitrogen*

X	Isomer Ratio			Total Rate Ratio	Apparent Partial Rate Factors			Ref.
	o	m	p	$\frac{X}{H}K$	F_o	F_m	F_p	
CH_3	65.4	2.4	32.2	1.86	3.65	0.13	3.59	51)
OCH_3	55.5	1.2	43.3	2.54	4.23	0.09	6.60	51)
Cl	57.4	0.9	4.17	0.44	0.76	0.01	1.10	51)
CO_2CH_3	64.3	34.4	1.3	0.30	0.58	0.31	0.02	19,36)
CN	68.9	31.1	0.0	—	—	—	—	19,36)
CF_3	53.4	45.6	1.0	—	—	—	—	36)

The values of the total rate ratios, taken on their own, would now support the idea of an attacking highly reactive, highly electrophilic triplet nitrene *or* of a highly reactive electrophilic singlet species. The apparent partial rate factors are, however, inconsistent with a rate-determining formation of a σ-complex by either species, and it was suggested that the above results, indicative of both high reactivity *and* high positional selectivity, were more in accord with a rate-determining addition of the initially formed singlet sulphonyl nitrene to give *N*-sulphonylaziridine intermediates *(42)* which underwent relatively fast ring opening to dipolar species *(43)* followed by (or concerted with) proton migration to give the final products *(43)* [51]. The relative reactivity of the aromatic nucleus ($\frac{X}{H}K$) would then reflect the influence of the

substituent upon the overall π-electron density of the nucleus. The substituent would determine the *direction* of ring opening in a product-, but not rate-, determining step. Clearly, the $+M$ substituent would favour opening according to path *a*. A likely potential energy diagram was given for this process. Consequently, the "partial rate factors" above are without meaning as a measure of the influence of the substituent upon the nuclear reactivity at individual positions since the rate-determining step is not the attack at these positions. Formation of an aziridine by addition to a benzene "double bond" rather than direct substitution was also suggested [61] to explain the predominant formation of 1-, rather than 9-, substituted product from benzenesulphonyl azide and anthracene.

p-Xylene was found to be twice as reactive as benzene towards tosyl azide, and a benzene "double bond" eight times more reactive towards singlet sulphonyl nitrene than a carbon-hydrogen bond in cyclohexane [8,12].

One surprising feature emerged from the reactions of methanesulphonyl azides with aromatic solvents at 120 °C, and that was the total absence of any *N*-sulphonylazepine derivatives *(45)* from the reaction products. The latter would have been expected from the electrocyclic rearrangement of *42* of the same type as had been observed in the reac-

tions with ethyl azidoformate [62] and cyanogen azide [63]. No *(45)* could be detected even by thin layer chromatography. The explanation for this has now been given [38]. An equilibrium exists between the aziridine

46, azepine *47*, and the dipolar intermediate *48*: at 120 °C, the temperature normally used in the thermolysis, it lies almost completely on the side of *48* which rearranges irreversibly to anilide *49*, and no azepine can be detected (thermodynamic control). Formation of azepine *47* is the kinetically controlled pathway, however, and if a trapping agent (e.g. tetracyanoethylene) is introduced into the reaction mixture prior to the decomposition, the azepine may be trapped at the expense of the formation of *48* to give the adduct *50*. Thus, while no *47* can be detected at 120 °C, the small concentrations present at equilibrium can be trapped, thus destroying the reversibility of the process.

Substituted adducts similar to *50* have been obtained from the reactions carried out in chlorobenzene and in toluene [19]. Whereas methanesulphonyl azide does not thermolyze appreciably below 120 °C, when a solution of $CH_3SO_2N_3$ in benzene was heated at 80 °C for 100 hr, *47* (ca. 0.5%) could be detected by thin layer chromatography but no *49* [38]. Almost all the azide remained undecomposed. Similarly, very small, amounts of *47* were observed, together with much tar and undecomposed azide, on photolysis of $CH_3SO_2N_3$ in benzene at room temperature or at 80 °C [19]. This confirms that *azepine formation is the kinetically controlled process*, while the anilides are the products of thermodynamic control.

While the isomer ratios and reactivities in the reaction of methanesulphonyl nitrene with toluene, anisole, and chlorobenzene can be explained as above on the basis of a rate-determining addition of the singlet to give an aziridine intermediate, it would be expected that an electron-attracting substituent would not only deactivate the nucleus but would also direct opening of the aziridine preferentially to that dipolar intermediate which would give the *m*-isomer. Indeed, the proportion of *m*-isomer does increase markedly on going from toluene as substrate (2.4%) to benzonitrile (31.1%) and methyl benzoate (34.4%) (Table 2), but the *o*-isomer still predominated. With benzenesulphonyl azide, the *m*-isomer was the main product with methyl benzoate, and the only product reported with benzoyl chloride (Table 1). The yields of anilides also dropped appreciably in the methanesulphonamidations: 5.4% with C_6H_5CN and 21.4% with $C_6H_5CO_2Me$, but no product of addition to the substituent could be isolated. When nitrobenzene was the substrate, a dramatic change took place [36]. The results are summarized in Table 3.

$$MeSO_2N_3 + C_6H_5NO_2 \xrightarrow[-N_2]{\Delta}$$

NO$_2$ group structure with —NHSO$_2$Me + $C_6H_5NHSO_2Me$ + $C_6H_5OSO_2Me$

51   49   52

$$+ \; MeSO_2NH_2$$
53

Table 3. *Reaction of MeSO$_2$N$_3$ with C$_6$H$_5$NO$_2$* (%-products[1])

Reaction conditions	51	49	52	53
Degassed under N_2	5.3 ($o:m:p$=55.4:13.4:31.2)	18.6	2.5	11.7
Under pressure (some air present)	2.8 (mainly *ortho*)	15.2	2.1	10.9
Open to air	2.0	4.5	2.6	9.9
O_2 bubbled through	0.0	0.3	2.5	7.5

[1]) Much tar was always formed.

These results are best interpreted in terms of a rate-determining substitution by a highly electrophilic radical. Thus, the isomer ratio obtained using degassed nitrobenzene is similar to that observed in the homolytic *p*-nitrophenylation of nitrobenzene ($o:m:p = 58:15:27$) [64]. The displacement of a nitro group by radicals has been reported. For example, hydroxyl radicals and $C_6H_5NO_2$ give some phenol [65]. Displacement of a nitro group was observed by Hey and Mulley in a Pschorr-cyclization [66]. Evolution of nitric oxide in the decomposition of sulphonyl azides in nitrobenzene has been observed but not accounted for [7,54]. The effect of oxygen upon the yields of *51* and *49* is consistent with the interception of the triplet diradical by oxygen before it can react with nitrobenzene. On the other hand, the yield of the sulphonate ester was unaffected by the presence of oxygen. Since, as discussed earlier,

25

sulphonyl azides can undergo S—N bond cleavage in the presence of radicals [8] the formation of 52 can be explained as follows:

$$MeSO_2N_3 + R\cdot \longrightarrow MeSO_2\cdot + RN_3$$

$$MeSO_2\cdot \xrightarrow{[O]} MeSO_3\cdot \xrightarrow{C_6H_5NO_2} MeSO_3C_6H_5 + [NO_2\cdot]$$

There is a parallel for oxygen abstraction by a sulphonyl radical either from nitrobenzene or by disproportionation: thermolysis of benzenesulphonyldiazomethane (54) in benzene gives, among other products, some sulphonate ester (55) [67], which has been formulated as follows:

$$C_6H_5SO_2CHN_2 \xrightarrow{\Delta} C_6H_5SO_2CH \xrightarrow{RH} C_6H_5SO_2CH_2\cdot$$
$$\downarrow \qquad\qquad \left. \begin{array}{l} \end{array}\right] \rightarrow C_6H_5SO_3CH_2SO_2C_6H_5$$
$$54 \qquad C_6H_5SO_2\cdot \xrightarrow{[O]} C_6H_5SO_3\cdot$$
$$55$$

Thus, the thermolysis of sulphonyl azides in aromatic solvents gives the singlet nitrene. If the surrounding molecules are sufficiently reactive this will add immediately to form an aziridine intermediate. If, on the other hand, the aromatic substrate is unreactive towards electrophilic addition (e. g. nitrobenzene), the singlet nitrene has time to drop to the ground state triplet (alternatively, but less likely, the substituent could perhaps catalyze the singlet-triplet conversion) and the pattern of attack now observed is typical of an electrophilic radical. With aromatic substrates that are not as deactivated towards electrophilic attack as is nitrobenzene (PhCO$_2$Me, PhCN, PhCF$_3$), this would occur only to a more limited extent and the pattern of substitution would then be in accord with an attack by both species, the singlet accounting for the marked increase in the proportion of m-isomer, but *ortho* still predominating because of the triplet contribution. Some additional support for the mixed mechanism comes from the fact that a plot of log $\frac{X}{H}K$ vs. σ_p (based, admittedly, on a very limited number of values available) is linear but the point for $X = p\text{-CO}_2Me$ falls considerably off the line.

In the hope of catalyzing the single → triplet conversion and then observing the pattern of substitution of the latter, reactions were carried out between MeSO$_2$N$_3$ and C$_6$H$_5$CO$_2$Me and C$_6$H$_5$CF$_3$ in the presence of suitable additives. Some of the results are summarized in Table 4 [36]. What was observed was a drop in the yield of substitution products and an increase in the proportion of m-isomer which thus became the predominant isomer, as expected of the direction of opening of an aziridine ring (42) when an electron-attracting substituent is present. The results

Table 4. *Methanesulphonamidation of $C_6H_5CO_2Me$ and $C_6H_5CF_3$ in the presence of various additives* [36]

Conditions	o	m	p	%$MeSO_2NHC_6H_4X$	%$MeSO_2NH_2$
			(a)		
			$PhCO_2Me$		
Sealed tubes under N_2	64.3	34.4	1.3	21.4	4.6
With oxygen	55.1	42.6	2.3	20.5	5.3
CCl_4 (40 molar excess)	62.9	35.0	2.1	12.9	[1]
CH_2Br_2 (40 molar excess)	29.9	57.5	12.6	1.1	44.2
Co^{III}acetyl-acetonate	32.3	63.3	4.4	2.2	[1]
Mn^{II}acetyl-acetonate	27.3	68.2	4.5	1.3	[1]
Mn^{II}acetyl-acetonate (trace)	61.4	37.1	1.5	16.6	[1]
$MnCl_2 \cdot 4\,H_2O$	61.6	36.7	1.67	13.6	[1]
Gattermann Copper	56.2	40.5	3.3	5.8	[1]
Iron powder	60.7	37.9	1.4	19.1	[1]
			(b)		
			$PhCF_3$		
Degassed under N_2	53.4	45.6	1.0	20.4	21.9
With oxygen	48.0	47.5	4.5	24.4	16.0
CH_2Br_2 (20 molar excess)	34.3	50.0	15.7	0.94	46.5
Cu^{II}acetyl-acetonate	38.2	58.1	3.7	4.3	29.5
Mn^{II}acetyl-acetonate	43.9	54.1	2.0	4.1	29.0
$Co_2(CO)_8$	31.2	66.8	2.0	2.9	16.1
$Fe_3(CO)_{12}$	30.7	64.0	5.3	0.75	61.5
$Fe(CO)_5$	23.8	64.7	6.5	0.55	53.2

[1] Not determined.

appear to indicate a trapping to a certain extent by the additive (or a side-tracking to an intermediate that hydrogen-abstracts) of the triplet species present or formed, so that the pattern of substitution becomes more characteristic of an attack by a singlet species the more efficient the trapping becomes. With CH_2Br_2 present in excess, the main product was that of hydrogen-abstraction (presumably by triplet nitrene) from CH_2Br_2 and the yield of substitution-product dropped considerably. CCl_4 was ineffective. Transition metal compounds had a similar effect to CH_2Br_2 but to different extents, depending on the nature of the addendum and its concentration.

In this connection it may be of interest to mention that a wide variety of reagents such as the fatty acid salts of Cu^{II}, Mn, Co, Ni, and Al, and magnesium acetylacetonate were reported to have no effect upon the rate of decomposition of *n*-octadecyl azidoformate in diphenyl ether solution [58].

The thermal, but not the photochemical, decomposition of ferrocenylsulphonyl azide *(14)* in benzene gave some intermolecular aromatic substitution product $FcSO_2NHC_6H_5$ (6.5%) but no intermolecular cyclization product *(17)*. Contrariwise, photolysis of *14* in benzene gave *17* but no anilide [17].

3.4. Addition to Olefins

Addition of carbethoxynitrenes to olefinic double bonds occurs readily. Addition of both the singlet and the triplet species can take place, the former stereospecifically, the latter not [49]. Additions of sulphonyl nitrenes to double bonds have not been demonstrated except in two instances in which metals were present. The reason is that either addition of the starting sulphonyl azide to the double bond occurs to give a triazoline that loses nitrogen and yields the same aziridine as would have been obtained by the direct addition of the nitrene to the olefin, or the double bond participates in the nitrogen elimination and a free nitrene is never involved [68]. The copper-catalyzed decomposition of benzenesulphonyl azide in cyclohexene did give the aziridine *56* (15%), which was formulated as an attack by the sulphonyl nitrene-copper complex on the double bond [24].

$$PhSO_2\ddot{N}{=\!\!=}Cu \; + \; \bigcirc\!\!| \longrightarrow \bigcirc\!\!\!\!\diagdown_{NSO_2PH} \; + \; \bigcirc\!\!\!\!\diagup^{NHSO_2PH}$$

56

Photolysis of ferrocenylsulphonyl azide in cyclohexene gave the corresponding aziridine derivative (9%), but thermolysis did not [17]. This could be an addition of the triplet nitrene to the olefin and studies on the stereospecificity of this reaction are under way.

3.5. Trapping by Nucleophiles

This has been mentioned at various points in this paper and may involve either a direct acid-base reaction of nitrene and nucleophile or, in some instances, reaction of the nitrene precursor with the nucleophile (or 1,3-dipolarophile) followed by loss of nitrogen. For example, the reaction of benzenesulphonyl azide with pyridine to give *31* (Ar = Ph) [69] could either involve a free nitrene or a concerted process in which the lone pair on the pyridine nitrogen atom assists the elimination of molecular nitrogen. That some free nitrene can be involved in these reactions is clear from the isolation of some 3-benzenesulphonamido-2,6-lutidine

(57) from the reaction of benzenesulphonyl azide with 2,6-lutidine [42].

3.6. Molecular Rearrangement

Until recently, sulphonyl azides were thought not to undergo Curtius-type rearrangements [Curtius' "starre" or 'rigid' azides [53,70]]. Rearrangements have now been observed on photolysis of benzenesulphonyl azide in protic solvents and upon base-induced decomposition of N-(p-nitrobenzenesulphonoxy)benzenesulphonamide *(26)* in alcohols and in aniline [20]. In addition to the O—H insertion product *(58)* (15%), methyl N-phenylsulphamate *(59)* (23%) was isolated from the photolysis of the azide in methanol. The rearrangement of the trimethylammonium salt of *26* has already been discussed (Section 2.2). Both of these rearrangements probably involve protonated species and no free nitrenes are formed.

$$C_6H_5SO_2N_3 + CH_3OH \xrightarrow{h\nu} N_2 + \underset{58}{C_6H_5SO_2NHOCH_3} + \underset{59}{C_6H_5NHSO_3CH_3}$$

$$+ C_6H_5SO_2NH_2 + C_6H_5SO_3NH_4$$

The vapour-phase pyrolysis of benzenesulphonyl azide at 625 °C gave a 17.5% yield of azobenzene [71], and a trace of the latter was also reported when the azide was heated in boiling cyclohexanone [9].

When mesitylene-2-sulphonyl azide *(3)* is heated to 150 °C in *n*-dodecane, a Curtius-type rearrangement of the nitrene *(4)* occurs as discussed in Section 2.1 i to give 2,4,6-trimethylaniline and the hexamethylazobenzene [14]. A similar result has now been observed by a careful analysis of the thermolysis products of durene-3-sulphonyl azide in *n*-dodecane at 150 °C. The amine is definitely formed but the azo-compound could barely be detected [13].

4. Applications

Disulphonyl azides have been used as *cross-linking* and *foaming agents* for cellulose acetate [72] and a number other polymeric materials [73], and as *blowing agents* [77]. A number of patents have been granted on

$$R(SO_2N_3)_2 \; + \; \begin{matrix} -\!\!\!\!\left(CH_2\!-\!CH_2\right)\!\!\overline{}_n \\[1em] -\!\!\!\!\left(CH_2\!-\!CH_2\right)\!\!\overline{}_m \end{matrix} \quad \longrightarrow \quad \begin{matrix} -\!\!\!\!\left(CH\!-\!CH_2\right)\!\!\!- \\ | \\ NH \\ | \\ SO_2\!-\!R\!-\!SO_2\!-\!NH \\ | \\ -\!\!\!\!\left(CH_2\!-\!CH\right)\!\!\!- \end{matrix}$$

applications making use of the ability of sulphonyl nitrenes to insert into aliphatic C—H bonds. Thus, mono- and poly-(sulphonyl) azides have been used for the modification or the cross-linking of hydrocarbon polymers such as polypropylene and polyisobutylene [75] and polyvinylethers [76] and chlorides [77]. Other examples are also known.

Aromatic substitution by sulphonyl azides has been applied to the *synthesis of cyclic sulphonamides* not as readily available by other methods [16]. For example, thermolysis of biphenyl-2-sulphonyl azide *(60)* in *n*-dodecane [16] or in cyclohexane [78] at 150 °C gives high yields (particularly in the latter solvent) of the known [79] 6H-dibenzo[c,e][1,2]thiazine 5,5-dioxide (61). Thermolysis of 2-phenoxybenzenesulphonyl azide (62)

in n-dodecane at 135 °C gave the cyclic sulphonamide 63 (15%) together with other products discussed earlier [16]. In the thermolysis of diphenyl sulphide 2-sulphonyl azide (64) the intermediate singlet nitrene is trapped by the sulphide sulphur to give the dithiazole 65. It is conceiv-

able that 65 could result from a concerted attack by sulphur at the azide and nitrogen elimination.

The intramolecular insertion of a sulphonyl nitrene into a side-chain methyl group to give 36 and 37 has already been mentioned, as has the intramolecular cyclization of ferrocenylsulphonyl azide to give the bridged ferrocene derivative (17).

Acknowledgements

The support of this work by N.S.F. grant GP-8869 and PHS grant NBS-8716 (to R.A.A.) and by grants from the National Research Council of Canada (to R.A.A. and R.G.S.) is gratefully acknowledged.

31

R. A. Abramovitch and R. G. Sutherland

5. References

1) Abramovitch, R. A., Davis, B. A.: Chem. Rev. *64*, 149 (1964). — Horner, L., Christmann, A.: Angew. Chem. *75*, 707 (1963).
2) Smolinsky, G., Wassermann, E., Yager, W. A.: J. Am. Chem. Soc. *84*, 3220 (1962).
3) — Snyder, L. C., Wasserman, E.: Rev. Mod. Phys. *35*, 576 (1963).
4) Moriarty, R. M., Rahman, M., King, G. J.: J. Am. Chem. Soc. *88*, 842 (1966).
5) Horner, L., Christmann, A.: Chem. Ber. *96*, 388 (1963).
6) Leffler, J. E., Tsuno, Y.: J. Org. Chem. *28*, 190 (1963).
7) — — J. Org. Chem. *28*, 902 (1963).
8) Breslow, D. S., Sloan, M. F., Newburg, N. R., Renfrow, N. B.: J. Am. Chem. Soc. *91*, 2273 (1969).
9) Balabanov, G. P., Dergunov, Y. I., Gal'perin, V. A.: J. Org. Chem. USSR *2*, 1797 (1966).
10) Takemoto, K., Fujita, R., Imoto, M.: Makromol. Chem. *112*, 116 (1968).
11) Balabanov, G. P., Dergunov, Y. I., Golov, V. G.: Zh. Fiz. Khim. *40*, 2171 (1966); Chem. Abstr. *65*, 19974 (1966).
12) Sloan, M. F., Breslow, D. S., Renfrow, W. B.: Tetrahedron Letters 2905 (1964).
13) Abramovitch, R. A., Holcomb, W. D.: Unpublished results.
14) — — Chem. Commun. 1928 (1969).
15) Leffler, J. E., Gibson, H. H.: J. Am. Chem. Soc. *90*, 417 (1968).
16) Abramovitch, R. A., Azogu, C. I., McMaster, I. T.: J. Am. Chem. Soc. *91*, 1219 (1969).
17) — — Sutherland, R. G.: Abstract G. 4, Fourth International Conference on Organometallic Chemistry, Bristol, England, July, 1969.
18) Shingaki, T.: Sci. Rep. Coll. Gen. Educ., Osaka Univ. *11*, 67, 81 (1963); Chem. Abstr. *60*, 6733, 6734 (1964).
19) Uma, V.: Ph. D. Thesis, University of Saskatchewan (1967).
20) Lwowski, W., Scheiffele, E.: J. Am. Chem. Soc. *87*, 4359 (1965).
21) Reagan, M. T., Nickon, A.: J. Am. Chem. Soc. *90*, 4096 (1968).
22) Hammond, G. S., Leermakers, P. A.: J. Am. Chem. Soc. *84*, 207 (1962).
23) Horner, L., Bauer, G.: Tetrahedron Letters 3573 (1966).
24) Abramovitch, R. A., Azogu, C. I., Sutherland, R. G.: Chem. Commun. 1439 (1969).
25) — — — Unpublished results.
26) Fry, A. J., Lui, R. S. H., Hammond, G. S.: J. Am. Chem. Soc. *88*, 4781 (1966).
27) Richards, J. H.: J. Paint Tech. *39*, 569 (1967).
28) Kirmse, W.: Carbene Chemistry. New York: Academic Press 1967.
29) Dekker, M., Knox, G. R.: Chem. Commun. 1243 (1967).
30) Doedens, R. J.: Chem. Commun. 1271 (1968).
31) Campbell, C. D., Rees, C. W.: Chem. Commun. 537 (1969).
32) Abramovitch, R. A., Davis, B. A.: J. Chem. Soc. (*C*) 119 (1968).
33) Kwart, H., Kahn, A. A.: J. Am. Chem Soc. *89*, 1950 (1967).
34) — — J. Am. Chem. Soc. *89*, 1951 (1967).
35) Franz, J. F., Osuch, C.: Tetrahedron Letters 837 (1963).
36) Abramovitch, R. A., Knaus, G. N., Uma, V.: J. Am. Chem. Soc., *91*, 7532 (1969).
37) — — Unpublished results.
38) — Uma, V.: Chem. Commun. 797 (1968).
39) Okahara, M., Swern, D.: Tetrahedron Letters 3301 (1969).
40) Robson, P., Speakman, P. R. H.: J. Chem. Soc. (*B*) 463 (1968).
41) Streith, J., Cassal, J. A.: Tetrahedron Letters 4541 (1968).
42) Abramovitch, R. A., Takaya, T.: Unpublished results.
43) — — Chem. Commun. 1369 (1969).

[44] Breslow, D. S., Sloan, M. F.: Tetrahedron Letters 5349 (1968).

[45] Carr, D., Seden, T. P., Turner, R. W.: Tetrahedron Letters 477 (1969).

[46] Acott, B., Beckwith, A. L. J., Hassanali, A.: Australian J. Chem. *21*, 185 (1968).

[47] McFarland, J. W., Burkhardt III., W. A.: J. Org. Chem. *31*, 1903 (1966).

[48] Breslow, D. S., Edwards, E. I.: Tetrahedron Letters 2123 (1967).

[49] Lwowski, W.: Angew. Chem. Intern. Ed. Engl. *6*, 897 (1967). — Lwowski, W., Simson, J.: Abstr. 153rd Meeting ACS, April (1967), 0163. — Simson, J.: Thesis, Yale University (1967).

[50] — Johnson, R. L.: Tetrahedron Letters 891 (1967).

[51] Abramovitch, R. A., Roy, J., Uma, V.: Can. J. Chem. *43*, 3407 (1965).

[52] Pritzkow, W., Timm, D.: J. Prakt. Chem. [4] *32*, 178 (1966).

[53] Curtius, T.: J. Prakt. Chem. [2] *125*, 303 (1930) and papers that follow.

[54] Dermer, O. C., Edmison, M. T.: J. Am. Chem. Soc. *77*, 70 (1955).

[55] Heacock, J. F., Edmison, M. T.: J. Am. Chem. Soc. *82*, 3460 (1960).

[56] Gutsche, C. D., Bachman, G. L., Coffey, R. S.: Tetrahedron *18*, 617 (1962).

[57] Smolinsky, G., Feuer, B. I.: J. Org. Chem. *29*, 3097 (1964).

[58] Breslow, D. S., Prosser, T. J., Marcantonio, A. F., Genge, C. A.: J. Am. Chem. Soc. *89*, 2384 (1967).

[59] Williams, G. H.: Chem. Ind. (London) 1285 (1961).

[60] Abramovitch, R. A., Koleoso, O. A.: J. Chem. Soc. (B) 799 (1969).

[61] Tilney-Bassett, J. F.: J. Chem. Soc. 2517 (1962).

[62] Hafner, K., König, C.: Angew. Chem. Intern. Ed. Engl. *2*, 96 (1963). — Lwowski, W., Maricich, T. J., Mattingly, T. W., jun.: J. Am. Chem. Soc. *85*, 1200 (1963).

[63] Marsh, F. D., Simmons, H. E.: J. Am. Chem. Soc. *87*, 3529 (1965).

[64] Hambling, J. K., Hey, D. H., Williams, G. H.: J. Chem. Soc. 3782 (1960).

[65] Loeble, H., Stein, G., Weiss, J.: J. Chem. Soc. 2704 (1950).

[66] Hey, D. H., Mulley, R. D.: J. Chem. Soc. 2276 (1952).

[67] Abramovitch, R. A., Roy, J.: Chem. Commun. 542 (1965).

[68] Franz, J. E., Osuch, C., Dietrich, M. W.: J. Org. Chem. *29*, 2922 (1964). — Oehlschlager, A. C., Zalkow, L. H.: Chem. Commun. 5 (1966), and references cited therein.

[69] Curtius, T., Rissom, J.: J. Prakt. Chem. [2] *125*, 311 (1930). — Buchanan, G. L., Levine, R. M.: J. Chem. Soc. 2248 (1950).

[70] — Z. Angew. Chem. *27*, 213 (1914). — Bertho, A.: J. Prakt. Chem. [2] *120*, 89 (1929).

[71] Reichle, W. T.: Inorg. Chem. *3*, 402 (1964).

[72] Ott, J. B.: (to Monsanto), U. S. Patent 2,518,249 (1950).

[73] — (to Monsanto), U. S. Patent 2,532,241 (1950); 2,532,242 (1950); 2,532,243 (1950).

[74] Adams, F. H.: (to American Cyanamid), U. S. Patent 2,830,029 (1958). — Hardy, W. B., Adams, F. H.: (to American Cyanamid), U. S. Patent 2,863,866 (1958).

[75] Breslow, D. S., Spurlin, H. M.: (to Hercules), U. S. Patent 3,058,944 (1962). — Newburg, N. R.: (to Hercules), U. S. Patent 3,287,376 (1966).

[76] Breslow, D. S.: (to Hercules), U. S. Patent 3,058,957 (1962).

[77] Robinson, A. E.: (to Hercules), U. S. Patent 3,261,785 (1966). — Breslow, D. S.: (to Hercules), U. S. Patent 3,261,786 (1966).

[78] Abramovitch, R. A., Chellathurai, T.: Unpublished results.

[79] Ullman, E., Grosse, C.: Ber. *43*, 2694 (1910).

Received December 22, 1969

Some Aspects of the Chemistry of Highly Halogenated Arynes

Dr. H. Heaney*

Department of Chemistry, The University of Technology, Loughborough, Leicestershire, England

Contents

I. Introduction

The chemistry of dehydrobenzene, the parent aryne, has become well established during the past almost twenty years [1]. It is essentially the chemistry of a short lived (half-life *ca.* 10^{-4} sec.), and highly electrophilic intermediate. It reacts with a large number of nucleophiles, and undergoes cyclo-addition reactions with a wide variety of compounds. A number of observations have led us, and others, to concentrate our efforts on the tetrahalogenobenzynes. It seemed reasonable to predict that the presence of four electron withdrawing substituents on the aryne *(1)* would result in a significant increase in the electrophilicity compared with that of benzyne.

---

* Based on lectures given at the Universities at Braunschweig and Darmstadt in September 1969.

$$Z \overset{\displaystyle Z}{\underset{\displaystyle Z}{\bigotimes}} \|$$

(1)

An investigation of the chemistry of the tetrahalogenobenzynes seemed to us to be more amenable to immediate study than would the chemistry of, for example, tetracyano- or tetranitrobenzyne.

Pentafluorophenylmagnesium halides [2-4], and pentachlorophenyl-magnesium chloride [5], were known and were the analogues of the known benzyne precursors the o-halophenylmagnesium halides. Pentafluoro-phenyl-lithium was also known to eliminate lithium fluoride to generate tetrafluorobenzyne [6]. It was anticipated that pentachlorophenyl-lithium would similarly act as a precursor for tetrachlorobenzyne. During the course of the investigations now under review pentachlorophenyl-lithium has been prepared [7-9]. While it was known that pentafluoro-phenyllithium was more stable then o-fluorophenyl lithium [1-6], it was not thought that this would be a major disadvantage. It was already known that although the 6-halogeno-2-benzenesulphonyl-phenyl-lithium derivatives [10] were unusually stable they were aryne precursors.

Polyfluoroarenes were known to form charge-transfer complexes with electron-rich arenes [11,12], and it was therefore possible that the precursors of the tetrahalogenobenzynes would form charge-transfer complexes with for example aromatic compounds, and thus enable reactions with these compounds to be studied. In this respect benzyne, generated from Grignard- or organolithium-reagents, is not amenable to this type of study since the precursor is itself a strong nucleophile, and adds to benzyne [1,13-16]. It was hoped that the tetrahalogenobenzynes would be sufficiently electrophilic to make addition reactions involving organometallic precursors less important. Benzyne is frequently generated from anthranilic acid, and since the tetrahalogenoanthranilic acids were all compounds reported in the literature (F [17,18], Cl [19], Br [20], I [21]) they were immediate potential precursors for the tetrahalogenobenzynes, along with the Grignard- and organolithium reagents already mentioned.

2. Reactions with Organometallic Reagents

A considerable amount of work, particularly involving tetrafluoro-benzyne, has been carried out in this area. The reaction of n-butyl-lithium with bromopentafluorobenzene, in a mixture of ether and light-petroleum, was found to give rise to significant amounts of 2-bromonona-fluorobiphenyl [22]. A detailed study of this reaction has shown that 2-iodononafluorobiphenyl is produced starting with iodopentafluoro-benzene and 2,2′,3,3′,4,4′,5,5′,6-nonafluorobiphenyl is obtained using pentafluorobenzene [23,24]. The mechanism of these reactions evidently involves the conversion of a biphenyl-lithium (5) into the final product (6) as shown (X = Br, I, or H).

(2)   *(3)*   *(4)*

(5)   *(6)*

This mechanism has been checked further by carrying out the analogous reaction using *o*-dibromotetrafluorobenzene [25]. The loss of lithium halide from 2-bromotetrafluorophenyl-lithium can give rise either to tetrafluorobenzyne or to 3-bromotrifluorobenzyne. However, the preferred loss is of lithium fluoride and hence the isomeric tribromohepta-fluorobiphenyls (7) and (8) were isolated.

(7)   *(8)*

The organolithium-reagent (5) is itself capable of forming the aryne (9) which can add pentafluorophenyl-lithium. This was found to occur and gave the terphenyl derivatives (10) and (11) in the ratio of 1:9 [26].

(9)   *(10)*   *(11)*

The organolithium compound which gives rise to the compound *(11)* can itself eliminate lithium fluoride, and mass spectrometric and other evidence has shown that this reaction is repeated until all the possible eliminations and subsequent additions have taken place [27].

One of the most interesting reactions of this type involves the intramolecular addition of the organolithium derivative to the aryne *(13)* which is derived from the dilithio-compound *(12)* [28]. This leads to the remarkably stable organolithium compound *(14)* which reacts with water to form the expected heptafluorobiphenylene, and with bromine to form 1-bromoheptafluorobiphenylene.

3. Pyrolytic Reactions

The pyrolysis of a number of compounds at temperatures around 600—800° and at pressures of the order of 10^{-2} mm. has been shown to give rise to benzyne. These compounds include for example indanetrione [29], and phthalic anhydride [30-33]. The dimerisation of benzyne to yield biphenylene has been used preparatively [31-33], and the pyrolysis of tetrafluorophthalic anhydride [34], and tetrachlorophthalic anhydride [31-33], gave the corresponding octahalobiphenylenes. In the case of the pyrolysis of tetrachlorophthalic anhydride some hexachlorobenzene is also formed, while the pyrolysis of tetrabromophthalic anhydride results in the formation of hexabromobenzene but no octabromobiphenylene. The disproportionation of tetrabromobenzyne to form carbon and bromine is a function of the high temperature involved and, as we shall see later, both tetrabromo- and tetraiodo-benzyne behave normally in solution.

Octafluorobiphenylene *(15)* has been obtained recently, along with octafluorofluorenone *(16)*, by the pyrolysis of silver tetrafluorophthalate [35].

The pyrolyses of 2-halogenobenzoates proceed at temperatures around 250° and yield xanthones [1]. The pyrolysis of sodium pentafluorobenzoate thus yields octafluoroxanthone *(18)* [36,37]. Although these reactions are known to involve arynes precise mechanistic details are not known but may well involve the following scheme.

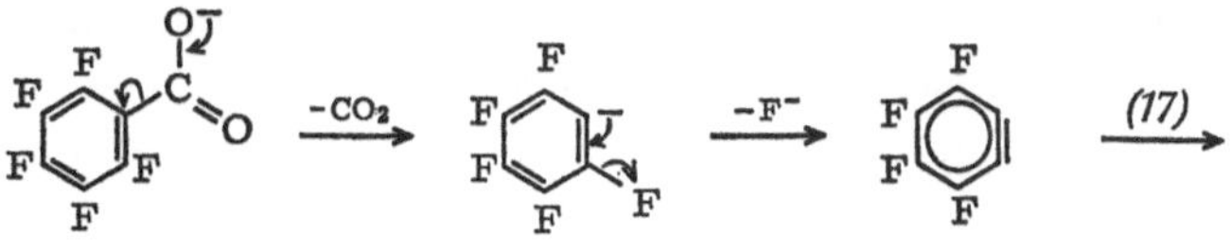

(17) → *(4)* → *(17)*

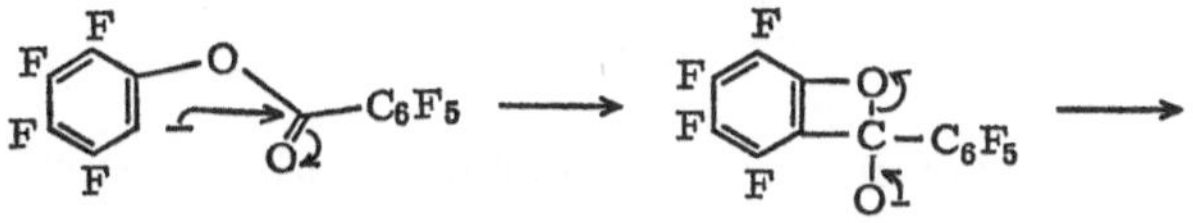

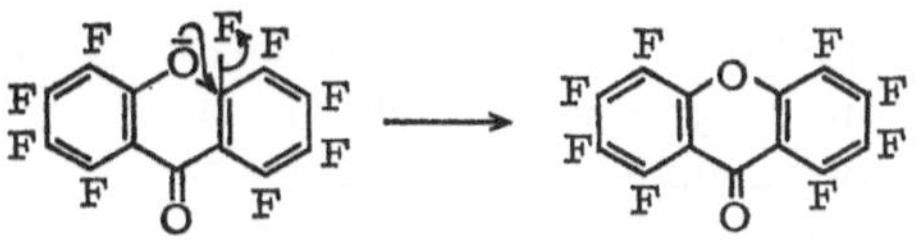

(18)

4. Reactions with Aromatic Hydrocarbons

Benzyne reacts with benzene to give a mixture of products in low yield. The original experiments [38] showed that the 1,4-cyclo-adduct (benzobarrelene) *(19)*, the valence-bond isomerised 1,2-cyclo-adduct (benzocyclo-octatetraene) *(20)*, and the product of insertion into a carbonhydrogen bond (biphenyl) *(21)*, were obtained in 2,8, and 6% yields respectively.

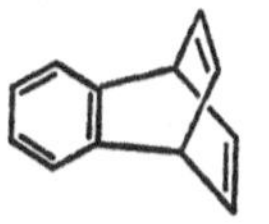

(19) 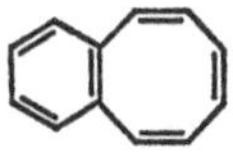   *(20)* 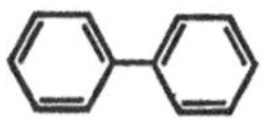   *(21)*

However, in a re-investigation of the reaction of benzyne with benzene [39], it was shown that the original method of isolating benzenediazonium-2-carboxylate resulted in the contamination of the zwitterion with silver salts [40,41]. In the absence of silver ions biphenylene, and the compounds *(19)*, *(20)*, and *(21)* were obtained in about 2, 17, 0.05, and 2% yields respectively. It was suggested [39] that a benzyne-silver complex is formed in the presence of the silver ions and that reactions of the complex, which is more electrophilic than benzyne, led to the increased yield of the compounds *(20)* and *(21)*. It is of interest that when benzyne was generated in benzene by the oxidation of 1-amino-benzotriazole *(22)* with lead tetra-acetate the only product detected was biphenylene, which was isolated in 83% yield [42].

$$\begin{array}{c} \text{(structure of 1-aminobenzotriazole)} \\ NH_2 \end{array}$$

(22)

During this period of time we, and a number of other research groups, have been investigating the reactions of highly halogenated arynes and hetarynes with aromatic hydrocarbons, for the reasons outlined in the introduction. In a reaction of pentafluorophenylmagnesium chloride with ethylene oxide in the presence of benzene, it was shown that, as well as β-pentafluorophenylethanol *(23)* a by-product of molecular formula $C_{12}H_6F_4$ was produced [43].

$$C_6F_5MgCl \xrightarrow[\text{b) } H^+]{\text{a) } \triangle, C_6H_6} C_6F_5 . CH_2 . CH_2 . OH + C_{12}H_6F_4$$

(23)

Unfortunately the structural formula assigned was incorrect. A re-interpretation of the spectral data [44], showed that the compound was 5,6,7,8-tetrafluoro-1,4-dihydro-1,4-ethenonaphthalene (tetrafluorobenzo-barrelene) *(24)*.

$$\begin{array}{c} F \\ F \\ F \\ F \end{array} \text{(structure of tetrafluorobenzobarrelene)}$$

(24)

The same compound has been obtained by a number of other research groups, using a number of different methods of generating tetrafluoro-benzyne. These include the loss of magnesium halide from pentafluoro-phenylmagnesium halides [44–46], the loss of lithium fluoride from penta-fluorophenyl-lithium [47,48], and the aprotic diazotisation of tetrafluoro-anthranilic acid [49].

When we prepared tetrafluorobenzobarrelene *(24)* from pentafluoro-phenylmagnesium bromide in the presence of benzene we isolated another compound which, from ^1H n.m.r., i.r., and u.v. evidence, was evidently closely related to the compound *(24)*.

Analysis and mass spectrometry showed it to have the molecular formula $C_{12}H_6BrF_3$. This could have been formed by the addition of magnesium bromide to tetrafluorobenzyne followed by the elimination of magnesium bromide-fluoride to give bromotrifluorobenzyne *(30)* and hence the compound *(31)*. Analysis of the ^{19}F n.m.r. spectrum and more particularly the preparation of *(31)* from o-bromotetrafluorophenyl-magnesium bromide and benzene confirmed the suggested mechanism. In this latter reaction the ratio of *(31)* to *(24)* was 99:1 [54].

(30)   *(31)*

The analogues of the other products formed in the reaction of benzyne with benzene, 2,3,4,5-tetrafluorobiphenyl *(25)*, octafluorobiphenylene, and tetrafluorobenzocyclo-octatetraene *(26)*, have not been detected in the reactions of tetrafluorobenzyne with benzene.

(25)   *(26)*   *(27)*

All of these three compounds are available for use as gas chromatographic standards. The compound *(25)* has been prepared by standard methods. The photolysis of tetrafluorobenzobarrelene *(24)* in the absence of a photosensitizer forms the compound *(26)* slowly, together with tetrafluorobenzosemibullvalene *(27)*. This latter compound *(27)* is produced more rapidly in the presence of a photosensitizer (acetone) [50].

The formation of the benzocyclo-octatetraene *(26)* undoubtedly involves a singlet species which undergoes benzo-vinyl bridging rather than the vinyl-vinyl bridging of the triplet which leads, eventually to compound *(27)* [51,52]. In accord with this the photo-isomerisation of the adduct *(28)* formed by the reaction of tetrafluorobenzyne with cyclohexa-1,3-diene, results in the formation of tetrafluorobenzo-dihydrosemibull-valene *(29)* [53].

(28)   *(29)*

In view of the good yields obtained in these reactions we, and others, extended these studies to include the reactions of tetrafluorobenzyne [54-56], tetrachloro- [57], tetrabromo- [58,59], and tetraiodo-benzyne [59], with benzene and other substituted benzenes.

We were able to prepare pentachlorophenylmagnesium chloride [5], pentachlorophenyl-lithium [7-9], and tetrachloroanthranilic acid [19], relatively easily. However, we found that pentabromophenyl-lithium could not be formed in an acceptable yield from hexabromobenzene and that tetrabromoanthranilic acid was not readily available by the published methods. The preparation of tetrabromoanthranilic acid, in high yield (>85%), has now been achieved, and we find that all of the tetra-halogenbenzynes react with benzene with the exclusive formation of the 1,4-cyclo-adduct in good yield. The adduct of tetrabromobenzyne with benzene has been prepared very recently in 3% yield from pentabromo-phenyl-lithium [58].

X = F,   35%   ref. [49]
X = Cl, 55%   ref. [57]
X = Br, 67%   ref. [59]
X = I,   25%   ref. [59]

The pentafluorophenyl-group has an electronegativity value which has been estimated as being between that of chlorine and bromine [60-62]. As would be expected, 3-pentafluorophenyltrifluorobenzyne *(9)* reacts with benzene to form the 1,4-cycloadduct in good yield [63].

Before completing our discussion of the reactions of the tetrahalogen-obenzynes with benzene we should discuss the mechanisms by which the product could arise. Benzyne, in the majority of its reactions with both cyclic and acyclic dienes forms predominantly, or exclusively the 1,4-cyclo-adduct [1,64,65]. 1,2-Cyclo-addition reactions have also been observed, and when allylic hydrogen atoms are present products derived by the 'ene' reaction [66] are obtained [1]. Recent calculations predict that o-dehydrobenzene is a ground-state singlet species with the two electrons of the "third-bond" in the lower symmetric orbital [67]. Thus the thermal cyclo-addition reactions of benzyne with 1,3-dienes is an allowed concerted process by the conservation of orbital symmetry [68]. The expected stereospecificity has been observed in the reactions of benzyne with, for example, trans-trans-hexa-2,4-diene [64,65]. Orbital symmetry analysis of various modes of cyclo-addition to the benzene ring shows that thermal 1,4cyclo-addition reactions are orbital symmetry controlled, but photochemical cyclo-addition reactions are not [69]. It seems reasonable that the formation of the 1,4-cyclo-adducts of benzene occurs by an essentially concerted process.

We have also attempted to study the reactions of the tetrafluoro-o-phenylene di-radical (32) with benzene, by carrying out the photolysis of 1,2-di-iodotetrafluorobenzene in the presence of benzene. It is known that 1,2-di-iodoarenes give arynes on photolysis [70-72]. The only product derived immediately from o-dehydrotetrafluorobenzene was the 1,4-cyclo-adduct (24) which was, as expected, partially photoisomerised to (27) and a trace of (26) [73]. The photolysis resulted in the appearance of (24) and (27) before (26) was detected and hence we conclude that (26) was not a primary product.

It is unlikely that the compound (27) is derived directly from the reaction of an excited benzene with tetrafluorobenzyne even though the compound (27) is formally analogous to the photo-adducts formed by the irradiation of olefins in benzene [74,75]. A number of other products derived from the o-iodotetrafluorophenyl radical were also obtained [73]. These results suggest either that the tetrafluoro-o-phenylene di-radical (32) is identical with tetrafluorobenzyne or that if it is produced at a higher energy level it returns rapidly to the groundstate before it reacts with benzene. An alternative and perhaps more likely explanation is that the tetrafluorobenzyne formed arises by the concerted loss of both iodine atoms.

The reactions of the tetrahalogenobenzynes with mono-alkyl-benzenes result in the formation of both the possible 1,4-cyclo-adducts [54-56,59]. However, the ratio of the two adducts only approaches the statistical value in the reaction of tetrafluorobenzyne with t-butylbenzene; more of the adduct (34) is normally produced.

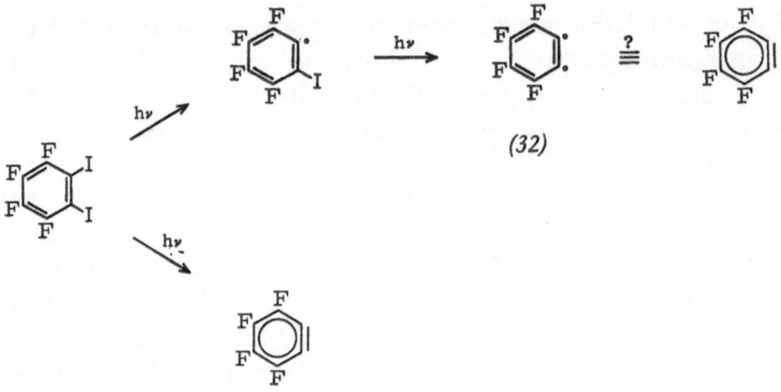

(32)

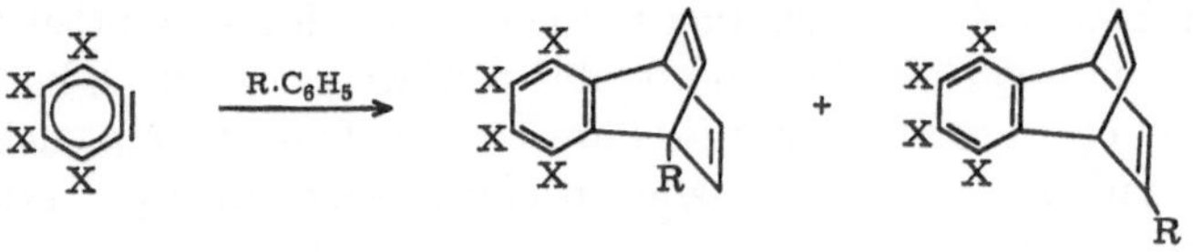

(33)   (34)

Ratio (33):(34) for X = F
R = Me 2.3:7.7
R = Et 2.9:7.1
R = Pr^i 3.1:6.9
R = Bu^t 3.2:6.8

These results suggest that the transition states leading to the formation of the cyclo-adducts (33) and (34) are product-like and that the
greater than statistical formation of adducts (34) is due to the increased
thermodynamic stability of a trisubstituted double bond. In agreement
with this explanation is the fact that in reactions with for example p-
xylene and durene (1,2,4,5-tetramethylbenzene) only the adducts (35)
and (36) were obtained [54-59]. Also as expected, two adducts were obtained with tetralin but only the compound (37) was obtained using
5,8-dimethyltetralin, which we may regard as a 1,2,3,4-tetra-alkylben-
zene [54].

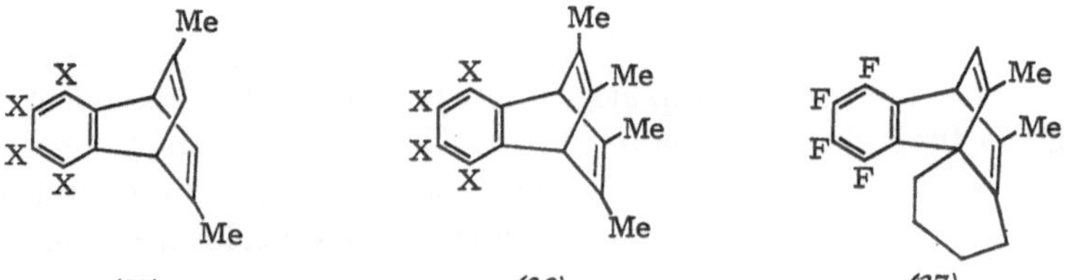

(35)   (36)   (37)

It is possible to force the reaction to give products bearing two alkyl-groups at bridgehead positions since 1,4-cyclo-adducts were obtained from hexamethylbenzene in 48% yield using tetrafluorobenzyne [54], and in 30% yield using tetrachlorobenzyne [57].

Since the adducts (33, X = F, R = Bu^t) and (34, X = F, R = Bu^t) were obtained in an almost statistical ratio one might conclude that there is no steric crowding in the transition state leading to the product (33, X = F, R = Bu^t). There is, on the other hand, a considerable steric effect in the product. The ^{1}H n.m.r. spectrum shows that there is a considerable barrier to rotation about the bridgehead carbon to t-butyl-group bond. The presence of long-range ^{19}F-^{1}H spin-spin coupling of proton containing substituents to the fluorine F* has proved to be extremely useful in structure determination, and in the case of the compound (38) a six-proton doublet ($|J|_{H-F} = 4.5$ Hz.) is observed downfield of a three proton singlet at 40° (Fig. 1). Two main conclusions can be tentatively drawn from this result and confirmed by variable tem-

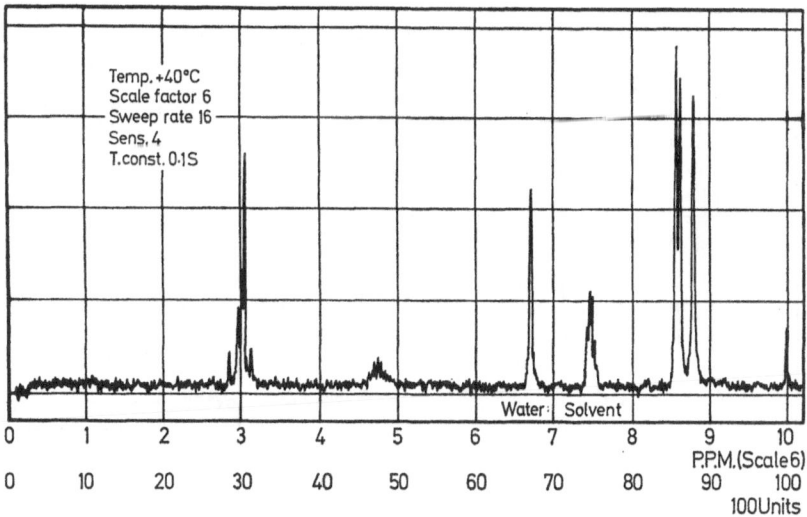

(38)

Fig. 1. ^{1}H n.m.r. spectrum of the Compound (38) in [^{2}H$_6$] Me$_2$SO at 40°

perature n.m.r. studies. The first conclusion is that the transmission of the spin-spin information is occurring by a through-space mechanism. The second conclusion is that a barrier to rotation favours the rotamer *(38)* at room temperature [76].

On raising the temperature (in hexadeuteriodimethylsulphoxide) the doublet and singlet signals collapse at about 135° (Fig. 2) and on raising the temperature a doublet (integral area nine protons) begins to appear (Fig. 3) and at 200° (Fig. 4), the maximum temperature attained, it is reasonably resolved and shows a coupling constant ($|J|_{H-F} = 2.9$ Hz.) which is approximately twothirds of the value obtained at 40°.

If the spin-spin information was being transmitted by the normal through-bond mechanism the upfield three proton signal would be expected to occur as a doublet because these protons are the only ones which can assume the required planar zig-zag conformation [77,78]. Preliminary results, using the change in chemical shift method [79], indicates that the energy barrier to rotation is of the order of 20 k.cal.mole [1]. As expected the silicon compound *(39)* shows a nine proton doublet

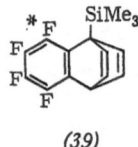

(39)

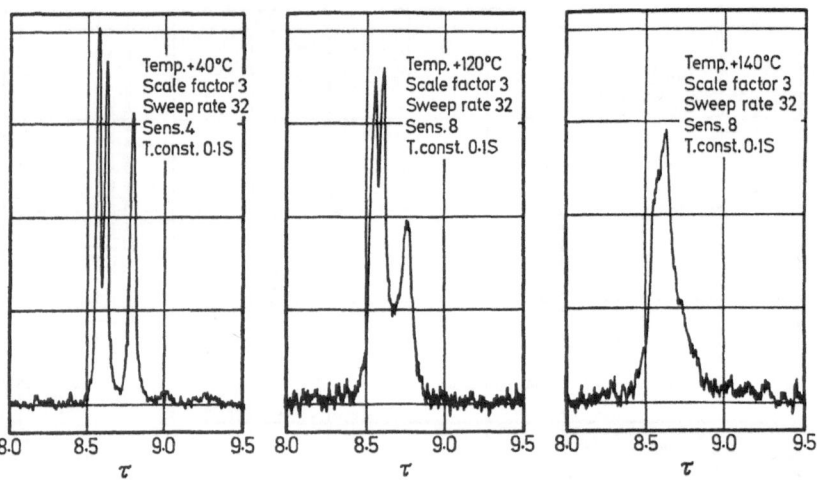

Fig. 2. Scale expansion of *t*-Butyl region of the n.m.r. spectrum at temperatures indicated

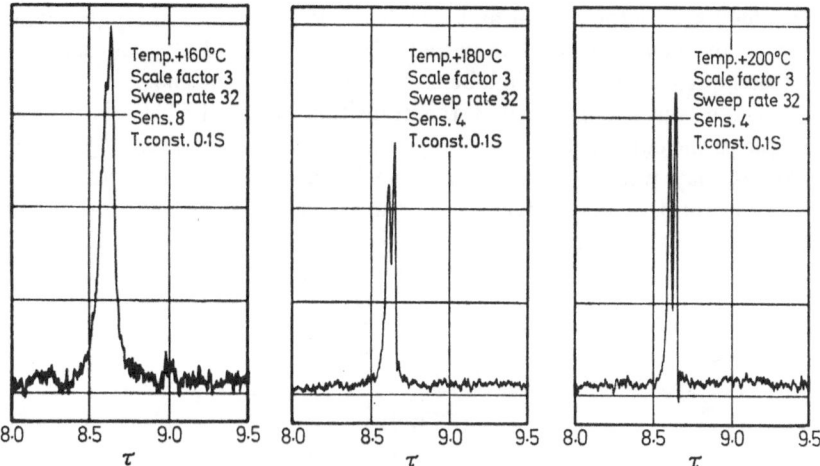

Fig. 3. Scale expansion of *t*-Butyl region of the n.m.r. spectrum at the temperatures indicated

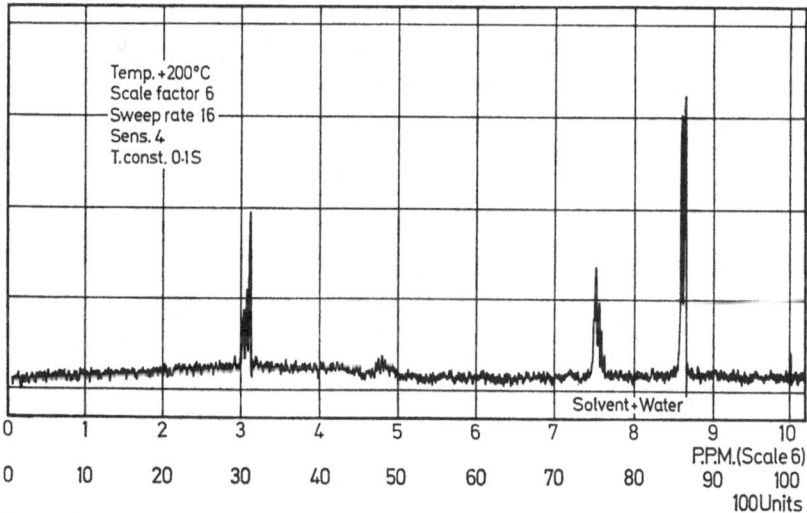

Fig. 4. ¹H n.m.r. spectrum of the Compound *(38)* at 200°

for the methyl protons [80]. Evidently the increase in the carbon to silicon bond length, compared to the analogous carbon to carbon bond-length results in the reduction of the steric effect.

In the analogous adducts *(33*, X = Cl or Br, R = Buᵗ) the steric hindrance to rotation is so severe that collapse of the *t*-butyl resonances, a

47

low-field six proton singlet and a higher-field three proton singlet, does not occur even at 200°. No adduct (*33*, X = I, R = Bu^t) was detected from the reaction of tetraiodobenzyne with *t*-butylbenzene.

These results suggest that the transition state which leads for example to the compound *(38)* is not symmetrical and completely product-like. It may well be that bond formation at the position *para* to the *t*-butyl group is more advanced in the transition state, as represented schematically in *(40)*.

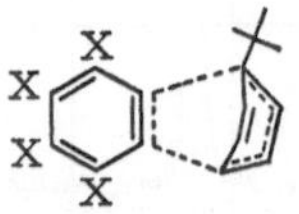

(40)

Another surprising feature appears in the reactions of certain alkyl-benzenes with tetrabromo- and tetraiodo-benzyne generated by the aprotic diazotisation of the corresponding anthranilic acids [59]. While only one product (*41*, X = Br or I, R = H or Me) was obtained in the reactions with benzene or *p*-xylene, three products, (*42*, X = Br or I, R = Me, Pr^i, or Bu^t), (*43*, X = Br or I, R = Me, Pr^i, or Bu^t) and a naphthalene derivative [a] were obtained in reactions with toluene, cumene (isopropylbenzene), and *t*-butylbenzene.

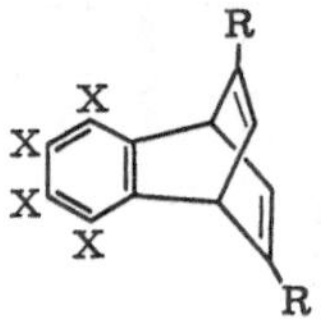

(41) 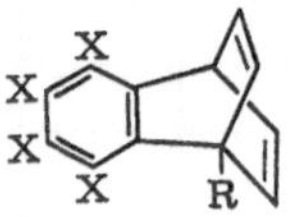   *(42)* 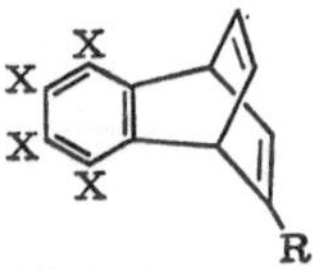   *(43)*

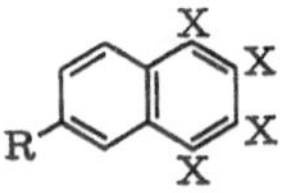

(44)

---

[a] The naphthalene derivatives are 6-substituted-1,2,3,4-tetrahalogenonaphthalenes (*44*, X = Br or I; R = Me, Pr^i, or Bu^t). We have shown that the compounds *(44)* are produced by excess of alkyl nitrite or more rapidly by an alkyl nitrite in the presence of a trace of acetic acid. Evidently this is a finely balanced reaction since it does not occur in the tetra-chloro-series under analogous conditions [81].

Since the nitrogen in pyridine is electron attracting it seemed reasonable to predict that the trihalopyridynes would also show the increased electrophilic character necessary to form adducts with aromatic hydrocarbons under similar conditions to those employed with the tetrahalogeno-benzynes. The availability of pentachloropyridine suggested to us and others that the reaction with n-butyl-lithium should lead to the formation of tetrachloro-4-pyridyl-lithium [82-84]. This has been achieved and adducts obtained, although this system is complicated by the ease with which pentachloropyridine undergoes nucleophilic substitution by tetrachloro-4-pyridyl lithium. Adducts of the type *(45)* have been isolated in modest yield both in the trichloro- and tribromo- [58] series.

(45)

X = Cl or Br
R = R′ = H; R = alkyl, R′ = H; R = R′ = Me

In general terms the yields of the adducts isolated in the reactions of the tetrahalogenobenzynes and the trihalogenopyridynes with arenes, were found to increase with increasing nucleophilicity of the arene. We have therefore carried out a number of competition reactions in which equimolar amounts of arenes were allowed to react with a small amount of tetrahalogenobenzyne. Thus with benzene and toluene tetrafluorobenzyne gives a ratio of 1:2.64 and with benzene and p-xylene a value of 1:6.7 was obtained [54]. Tetrachlorobenzyne, generated either from pentachlorophenyl-lithium or from tetrachloroanthranilic acid gives a ratio of *ca.* 1:20 [57]. Tetrabromobenzyne and tetraiodobenzyne give competition ratios of 1:18.5 and 1:19.5 respectively in reactions with equimolar amounts of benzene and p-xylene [59].

If the competition data are compared with electronegativity values for the halogens [85], then tetrafluorobenzyne is clearly in an anomalous position. The only reasonable explanation available at present is that tetrafluorobenzyne is so destabilized by the inductive effect of the fluorine atoms that it has lost a considerable amount of the selectivity which arynes normally show. Estimates for the heats of formation of the isomeric dichlorobenzynes and for tetrachlorobenzyne have recently been made from mass spectrometric studies and these do indicate a low stability for tetrachlorobenzyne [86]. Evidently more data are required for the tetrahalogenobenzynes.

The reactions of the tetrahalogenobenzynes with polycyclic aromatic compounds follow the expected paths. Thus with naphthalenes [54,57] cyclo-addition at the 1,4-positions occurs even with 1,4,5,8-tetramethyl-naphthalene [54]. Thus the tetrahalogenobenzynes are only able to destroy the aromaticity of one ring to form example the compound *(46)*.

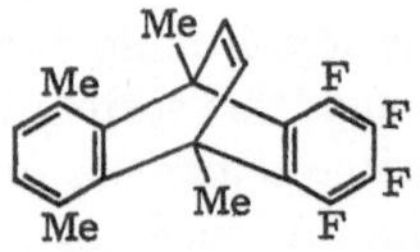

(46)

With anthracene, the formation of compounds *(47,* X = F or Cl) and *(48,* X = F or Cl) [54,57], parallels similar reactions of benzyne with anthracene, and substituted anthracenes [87-89].

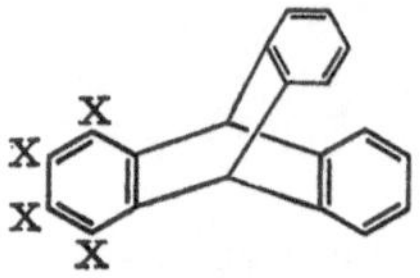

(47) 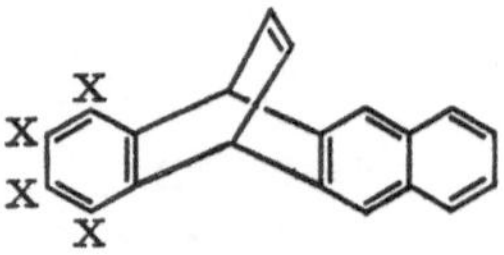   *(48)*

Whereas tetrafluorobenzyne only forms one adduct *(49)* with ace-naphthene, two adducts *(50)* and *(51)* were formed with acenaphthylene [90]. The formation of the compound *(51)* clearly reflects the high nucleo-philicity of the 1,2-double bond even though the cyclo-adduct cannot be formed by an orbital symmetry allowed concerted process.

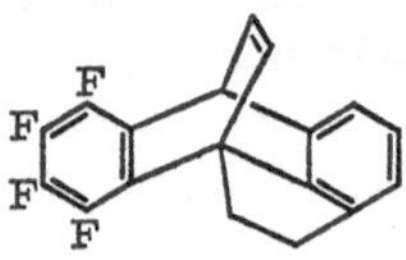

(49) 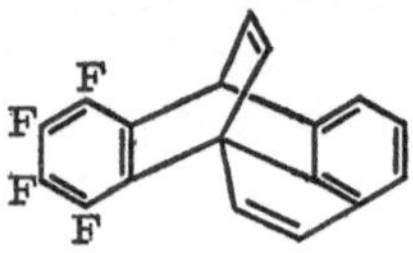   *(50)* 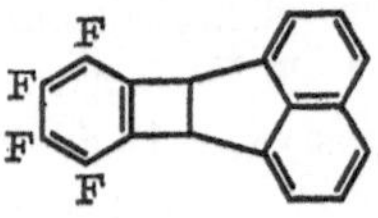   *(51)*

The reactions of tetrafluorobenzyne with 9,10-dihydrophenanthrene and phenanthrene yield the expected adducts formed by cyclo-addition at the 1,4-positions [90]. The reaction of tetrafluorobenzyne with 1,6-methanocyclodecapentaene *(52)* was carried out in order to study the mass spectral and thermal fragmentation of *(53)* [90]. In the event benzocyclopropene and 1,2,3,4-tetrafluoronaphthalene *(54)* were formed.

Attempts to form cyclo-adducts with biphenylene have interested a number of research groups. Charge-transfer complexes are formed by

(52)   (53)   (54)

biphenylene and polymethoxybiphenylenes with maleic anhydride [91], and tetracyanoethylene [92,93]. These do not collapse to form cyclo-adducts [91,93]. Similarly benzyne, generated from o-fluorophenylmagnesium bromide [15,16], or from 1-aminobenzotriazole [94], does not react with biphenylene. Molecular orbital calculations show the highest sum of the free valence index values for the 1,4-positions but the calculated product stabilities for the cycloaddition of maleic anhydride to biphenylene indicate that 2,4 a cyclo-addition should be favoured [95]. The tetrahalogenobenzynes do form the expected adducts (55, X = F or Cl) [96].

(55)

Although not strictly within the purview of the present section it is of interest to note that tetrafluorobenzyne forms the adducts (56) and (57) with [2.2]paracyclophane [97].

(56)   (57).

5. Reactions with Styrene and Substitution Styrenes

Several papers describing reactions of benzyne with substituted styrenes [98–100] had appeared before we [101] and a number of other groups [102, 103,56] began investigating the reactions of the tetrahalogenobenzynes with styrene and substituted styrenes. No well defined product had been obtained in the reaction of benzyne with styrene itself [104] although this reaction has now been investigated in some detail [105,106,101]. The reaction of benzyne with 2,4,6[^{2}H$_3$]styrene leads either to 9-phenyl-

9,10-dihydrophenanthrene *(59)*, by an 'ene' reaction of benzyne with the initial adduct *(58)*, or to the threo and erythroisomers of the compound *(60)* by the 'ene' reaction of *(58)* with styrene.

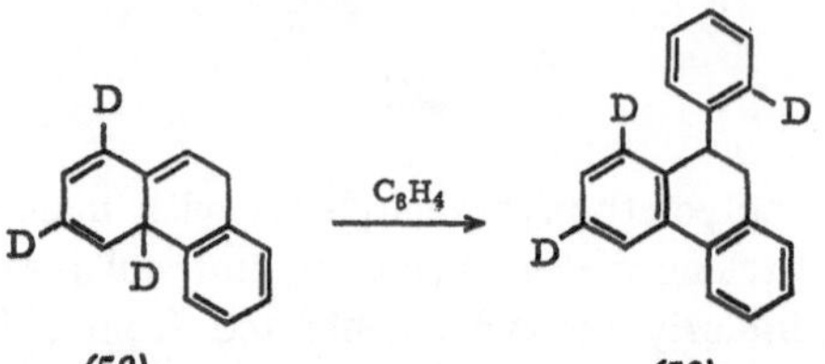

(58)   *(59)*

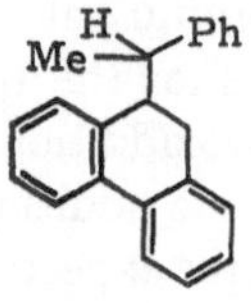

(60)

In our reactions we obtained either 1,2,3,4-tetrahalogeno-9,10-dihydrophenanthrenes or the related phenanthrenes or mixtures of these two types of compound. When styrene (5 mol.) was added to a solution of a solution of pentafluorophenyl-lithium in ether and the solution was allowed to warm to room temperature rapidly 1,2,3,4-tetrafluoro-9,10-dihydrophenanthrene *(61)* was obtained in good yield. The analogous product was obtained using pentachlorophenyl-lithium. When pentafluorophenyl-lithium was allowed to warm to room temperature slowly in the presence of styrene 1,2,3,4-tetrafluorophenanthrene *(62)* was obtained as the major product, together with some of the compound *(61)*. 1,2,3,4-Tetrachlorophenanthrene was the only product isolated when tetrachlorobenzyne was generated from tetrachloroanthranilic acid in the presence of styrene.

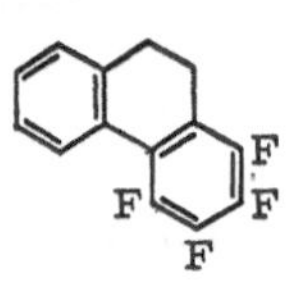

(61) 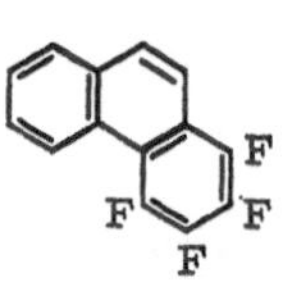   *(62)*

The reactions of a number of substituted styrenes have also been investigated. The majority give rise to analogues of either *(61)* or *(62)*. *trans*-β-Methylstyrene gave only 1,2,3,4-tetrafluoro-10-methylphenanthrene and no cyclo-adducts were obtained with β,β-dimethylstyrene. We assume that the steric requirement of the *cis*-β-methyl group prevents the formation of the transition state leading to the initial cyclo-adduct.

The stereochemistry of dienes has been found to have a pronounced effect in the concerted cyclo-additions with benzyne [64,65]. A concerted disrotatory cyclo-addition of tetrafluorobenzyne, leading for example with trans-β-methylstyrene to (63, R = Me), is likely and in accord with the conservation of orbital symmetry [68]. However while the electro-cyclic rearrangement of (63, R = H) to (65, R = H) is not allowed, base catalysed prototropic rearrangement is possible. A carbanion (64, R = H) cannot have more than a transient existence in the reaction of tetra-fluorobenzyne with styrene because no deuterium incorporation in (65) was detected when either the reaction mixture was quenched with deuterium oxide or when the reaction was conducted in the presence of a ten molar excess of deuteriopentafluorobenzene.

We therefore carried out a reaction of tetrafluorobenzyne with [2-^{2}H]styrene. The product obtained (65, R = H) was analysed by mass spectrometry and found to contain 90% [^{2}H$_1$], 5% [^{2}H$_2$], and 5% [^{2}H$_0$] which indicates a high intramolecularity.

(63)   (64)   (65)

A number of base catalysed intramolecular 1,3-proton transfer reactions have been observed previously and a 'conducted tour' mechanism proposed [108,109]. An analogous mechanism would account for the results of the labelling experiments in the reaction of tetrafluorobenzyne with styrene, in which pentafluorophenyl-lithium acts as the base. The highly intramolecular character in our reaction may result from the ability of the migrating conjugate acid (pentafluorobenzene) to exist as a charge-transfer complex with the extended π electron system containing negative charge. The failure to observe the chlorinated analogue in the reaction in which tetrachlorobenzyne was generated from tetrachloroanthranilic acid would therefore result from the absence of a suitable base which could act as a catalyst in the proton transfer. Similarly, although o-fluorophenylmagnesium bromide could remove the analogous proton in the benzyne reaction with styrene, fluorobenzene would not be capable of acting as the acid in the reprotonation step. In the presence of a large excess of styrene the intermediate (63, R = H) has been found to give an 'ene' product which is formally a 1:2 adduct of tetrafluorobenzyne and styrene [56].

If we consider the reactions in which a formal dehydrogenation occurs two experimental observations are significant. First, both α-methyl- and *trans*-β-methyl-styrene do yield the appropriate phenanthrene derivatives in reactions with tetrafluorobenzyne. Secondly, we showed that the 9,10-dihydrophenanthrenes are not dehydrogenated under the conditions used. Tetrafluorobenzyne was generated in the presence of styrene and 1,2,3,4-tetrafluoro-9,10-dihydro-6,8-dimethylphenanthrene which was unchanged at the end of the reaction even though the expected dihydro-phenanthrene and phenanthrenes were produced from styrene. We therefore carried out a reaction using *cis*-β-deuteriostyrene. The deuterium was not removed stereospecifically and we therefore concluded that the two hydrogen atoms were not removed from the intermediate *(66)* in a concerted process to form *(62)*. The most reasonable explanation of this result is that tetrafluorobenzyne abstracts a hydride ion from *(66)*. The removal of a hydride ion from the carbanion *(64)* is unlikely since the phenanthrene *(62)* was isolated in a significant amount in the reaction in which no deuterium incorporation was detected in *(61)* in the presence of deuteriopentafluorobenzene. Since pentafluorobenzene would be expected to form a charge transfer complex with the extended π-electron system of *(66)* it is not surprising that *(62)* was produced and this suggests that the hydride ion is removed from the methylene group rather than from the ring-junction methine group in the intermediate *(66)*. In addition the approach of tetrafluorobenzyne to remove a hydride ion from the ring-junction position would set up a favourable geometry for an 'ene' reaction, the transition state for which would presumably be of lower energy than that involving the removal of a hydride ion. The formation of the stabilised carbonium ion *(67)* could thus lead to the phenanthrene *(62)* by the loss of a proton.

Although we found that a 3-methyl substituent in a styrene inhibited cyclisation with tetrafluorobenzyne at position-2 reaction does occur with 2,3,4,5,6-pentafluorostyrene [102]. Dehydrofluorination of the first intermediate occurs to form 1,2,3,4,5,6,7,8-octafluorophenanthrene.

6. Reactions with Alkoxyaromatic Compounds

If the transition state leading to the formation of 1,4-cyclo-adducts is not necessarily symmetrical, one would anticipate that substituents on an arene which can release electrons by a conjugative mechanism would react with the tetrahalogenobenzynes to give non-statistical mixtures of adducts. Anisole reacts with the tetrahalogenobenzynes to give a mixture of the adducts *(68)* and *(69)*: the latter adducts actually being isolated as the ketones *(70)*. This result reflects the known feature of electrophilic substitution of anisole [110-113] and suggests that the transition state leading to *(68)* does not involve an absolutely synchronous formation of the two new bonds, as indicated in *(71)*.

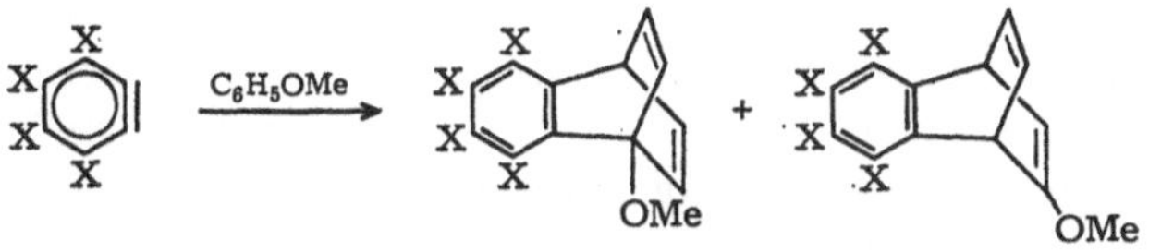

(68)   *(69)*

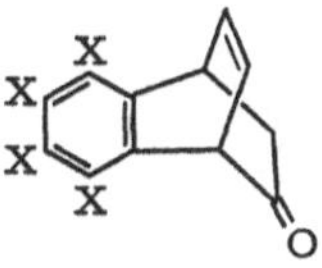

(70) 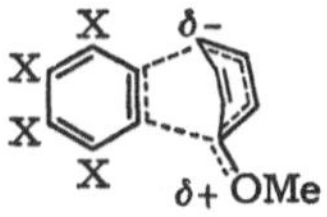   *(71)*

Ratio *(68)* : *(70)*
$X = F$ 4 : 1 (refs. [54,55])
$X = Cl$ 29 : 1 (ref. [59])
$X = Br$ 2.6 : 1 (ref. [59])

These results prompted us to investigate the reactions of tetrafluorobenzyne with other alkoxybenzenes [114] not only in order to attempt to explore mechanistic features more closely, but also because we envisaged that the ketonic products would undergo interesting reactions.

o-Dimethoxybenzene was found to give one major product *(72)* in moderately good yield and *m*-dimethoxybenzene gave only one product *(73)* in good yield.

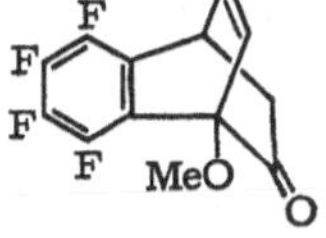

(72) 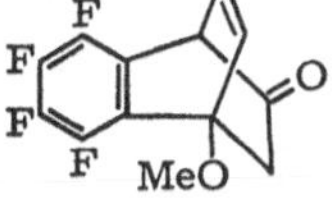   *(73)*

As we had hoped, *p*-dimethoxybenzene reacted with tetrafluorobenzyne to form initially a mixture of the two adducts *(74)* and *(75)* in the approximate ratio of 1:2.2. The compound *(65)* was again isolated as the diketone *(76)*. These results should be contrasted with our failure to detect an adduct anologous to *(74)* in reactions with *p*-xylene.

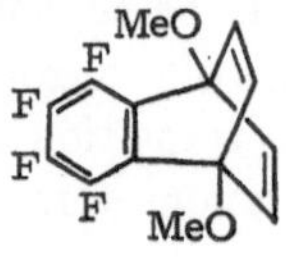

(74) 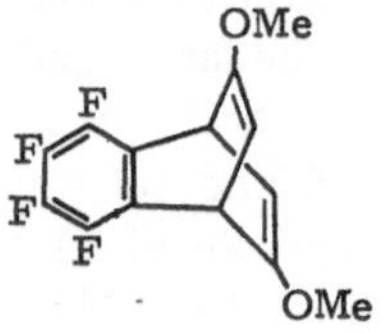   (75) 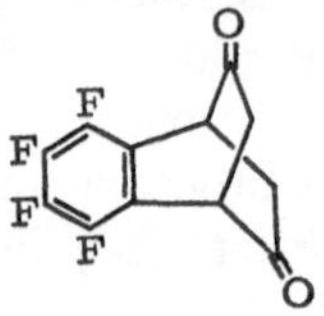   (76)

We have isolated enol-ethers in certain of our experiments, but n.m.r. and i.r. spectroscopy indicates that these are hydrolysed extremely rapidly. This may involve participation of the other double bond, involving for example, *(77)* initially.

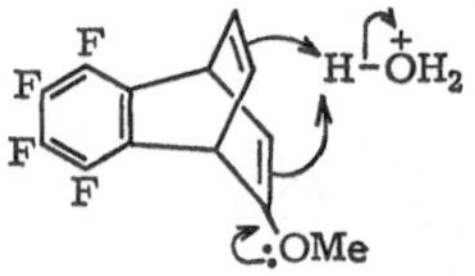

(77)

The reactions of benzyne with enol-ethers and enol-acetates have been much studied very recently [115–118]. We were not surprised therefore to isolate a product derived from the attack of tetrafluorobenzyne on the bis-enol-ether *(75)*. This product is derived from a $(2+2)$ π cyclo-addition and the available evidence suggests that this product has the structure *(78)*.

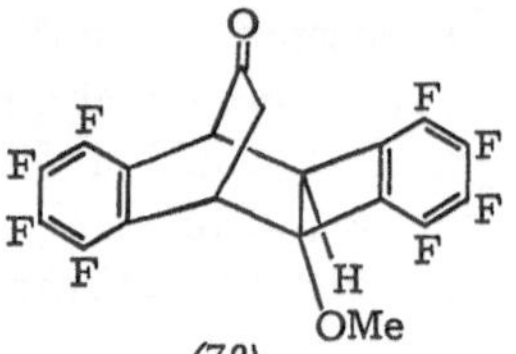

(78)

The mass spectral fragmentation of the adducts *(72)*, *(73)*, and *(78)* occurs by the loss of ketene from the molecular ion, and in accord with this and the results of other research groups [119–122], both *(72)* and *(73)* form ketene and *(79)* both thermally and photochemically.

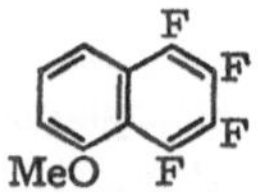

(79)

It was unfortunate that we did not detect any product derived from a diketone in the reaction of *m*-dimethoxybenzene with tetrafluorobenzyne. We therefore carried out a reaction of tetrafluorobenzyne with 1,3,5-trimethoxybenzene. The di-enol ether *(80)* could not be isolated, and after the removal of unreacted 1,3,5-trimethoxybenzene we isolated the phenolic acid *(81)* in good yield. This compound is undoubtedly formed by the hydrolysis of *(80)* followed by a retro-Claisen condensation, and aromatisation as shown below.

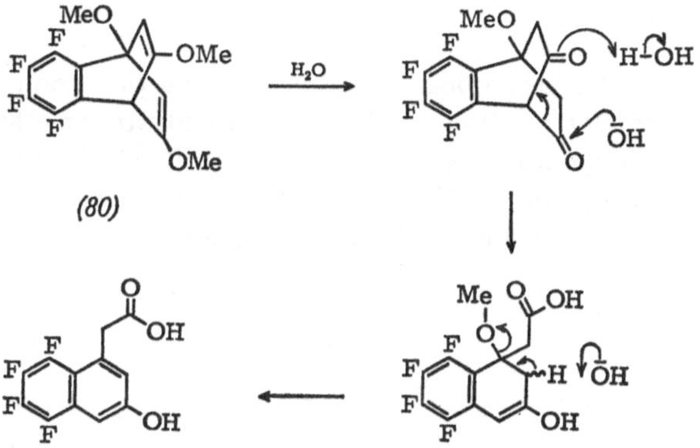

(80)

(81)

The reactions of the tetrahalogenobenzynes with alkoxynaphthalenes, in accord with the results previously mentioned, result in addition at the 1,4-positions. Thus tetrafluorobenzyne reacts with 1,2,3,4-tetrafluoro-5-methoxynaphthalene to form *(82)* in good yield [90]. Reactions with 2,3-dimethoxynaphthalene lead to the formation of the adducts *(83,* X = F or Cl) [123]. These compounds are relatively unstable and are transformed slowly in the open atmosphere. Thus *(83,* X = Cl) affords the di-ester *(84)*.

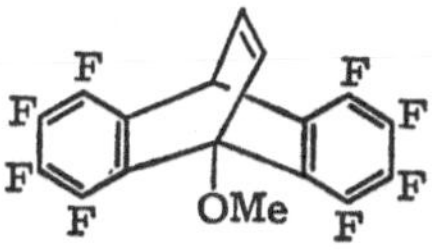

(82) 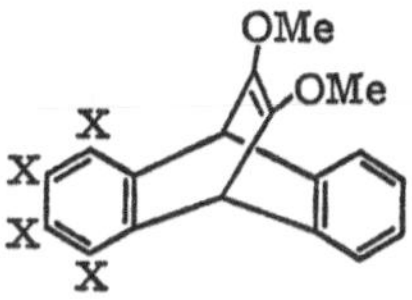   *(83)* 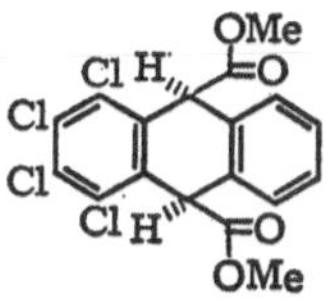   *(84)*

7. Reactions with Thioethers

The sulphur atoms in thio-ethers are more nucleophilic than the oxygen atoms in ethers. A number of reactions of benzyne with thio-ethers have been reported [1,124,125], and zwitter-ions are produced which are stabilised in a number of ways. It was not surprising, therefore, that no cyclo-addition reactions were observed in the reactions of tetrafluoro-benzyne and tetrachlorobenzyne with thioanisole. 1,2,3,4-Tetrahalogeno-5-phenylthiobenzenes (*85*, X = Cl or F) were isolated in good yield [126]. The mechanism by which the products arise was checked by means of labelling experiments. Thus the reaction of tetrafluorobenzyne with [*Me*-³H]thioanisole gave the compound (*85*, X = F) with 7.4% of the activity of the labelled thioanisole. The reaction of tetrafluorobenzyne with an equimolar mixture of [4-²H]phenyltrideuteriomethyl sulphide and thioanisole gave the compound (*85*, X = F) which was shown, by mass spectrometry, to contain only unlabelled and [²H₂] products. Thus the formation of the ylide (scheme below) involves an intramolecular proton-transfer.

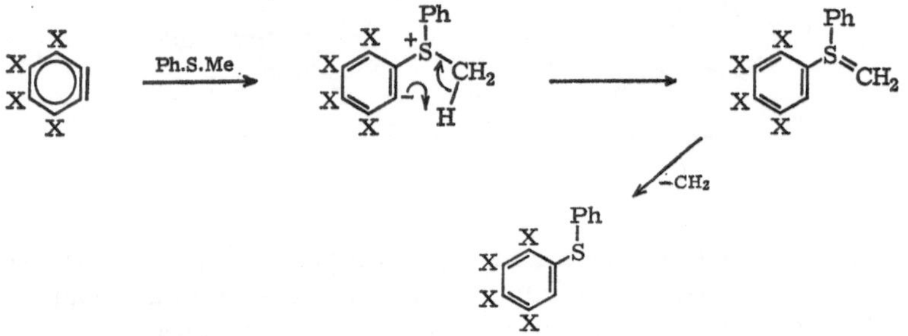

(85)

The reaction of tetrafluorobenzyne with pentafluorothiophenolate ion has been reported recently [127]. This reaction affords a particularly good yield of octafluorodibenzothiophen *(87)*. Evidently the anion *(86)* displaces fluoride ion particularly easily to form *(87)*.

(86)   *(87)*

The cleavage of thio-ethers suggested that the breakdown of the ylide derived from the reaction of an aryne with tetrahydrothiophen would be of interest. We found that tetrafluorobenzyne undergoes the expected reaction and that 2,3,4,5-tetrafluorophenylvinyl sulphide *(89)* was formed in good yield [128]. The electrocyclic elimination of ethylene from the ylide *(88)* parallels a number of other reactions which have been rationalised recently [129].

(88)   *(89)*

8. Reactions with Aliphatic Ethers

Benzyne is thought to interact with simple ethers such as diethyl ether to form zwitter-ions. However, simple products analogous to those obtained with for example diethyl sulphide have not been detected [1]. Apparently the more basic ether, 1,2-dimethoxyethane is cleaved by benzyne [130].

We have shown that a number of aliphatic ethers, containing sterically accessible β-hydrogen atoms, are cleaved in good yield when the tetrahalogenoanthranilic acids are diazotised in ethers [128,59]. Diethyl ether is cleaved to the tetrahalogenophenetole *(90)* and methyl-cyclohexyl ether affords the tetrahalogenoanisole. In this latter reaction we were able to detect cyclohexene and therefore a plausible, but as yet unproved, mechanism is as shown.

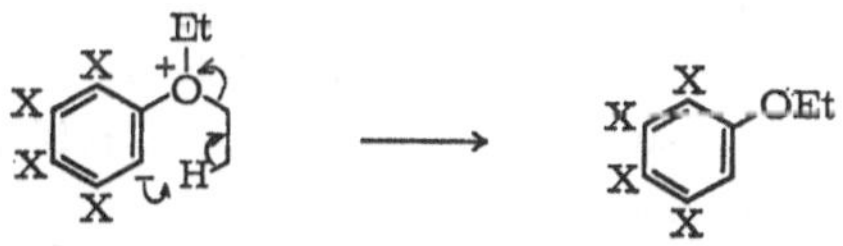

(90)

Little evidence for the production of simple products was obtained using tetrahydrofuran, but more complex products, for example the compound *(91)*, have been reported very recently from the reaction of benzyne with tetrahydrofuran [131].

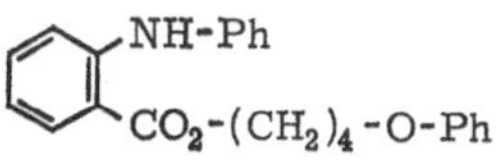

(91)

9. The Reaction of Tetrafluorobenzyne with *N,N*-Dimethylaniline

Products derived from intermediate zwitter-ions have been obtained in the reactions of benzynes with tertiary amines [1]. Benzyne generated by the reaction of *n*-butyl-lithium with fluorobenzene interacts with *N,N*-dimethylaniline to yield *N*-methyl diphenylamine and *N*-ethyl diphenylamine [132,133]. Using the reaction of chlorobenzene with *n*-butyl-lithium to generate benzyne resulted in the formation of increased amounts of 2-*N,N*-dimethylaminobiphenyl [132,133].

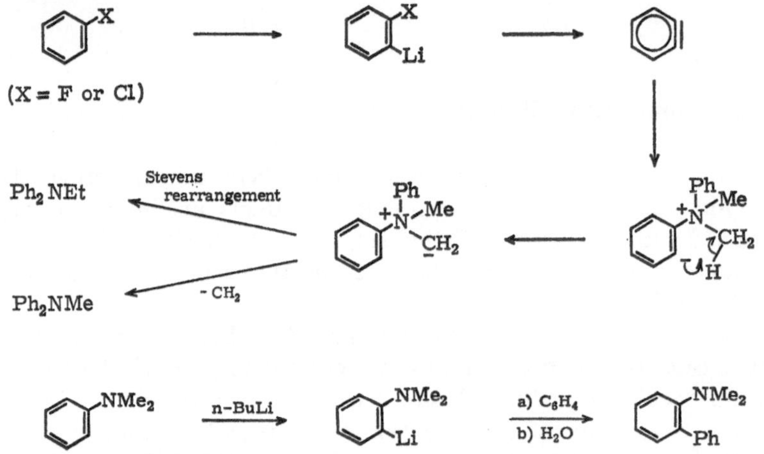
X
X
Li
(X = F or Cl)
Stevens rearrangement
Ph2 NEt
Ph2NMe
- CH2
Ph
Me
CH2
H
NMe2
n-BuLi
Li
a) C6H4
b) H2O
Ph

Five products were obtained in the reactions of tetrafluorobenzyne with *N,N*-dimethylaniline [134]. When we generated tetrafluorobenzyne by heating pentafluorophenylmagnesium chloride with an excess of *N,N*-dimethylaniline at 80° we obtained only three of the products. These were shown to result from the attack of tetrafluorobenzyne on the aromatic residue of the arylamine.

The major product, isolated in 10% yield, was shown to be the 1,4-cycloadduct *(92)* and the minor product, isolated in 0.75% yield, was the ketone (*70*, X = F) which evidently arises by the hydrolysis of the enamine *(93)*.

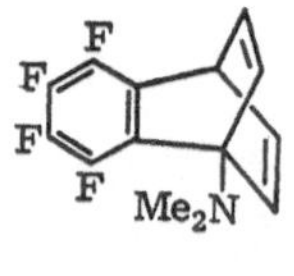
F
F
F
F
Me2N

(92)

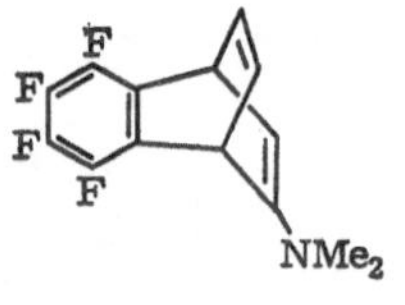
F
F
F
F
NMe2

(93)

The third product, which was formed in 1.5% yield, was found to be stable to brief heating with either dilute mineral acid or base, was not reduced by sodium borohydride in methanol, and did not have the properties of an enamine. It absorbed two moles of hydrogen in the presence of palladium to give a product which again failed to give reactions of an enamine. On the basis of this evidence and spectroscopic data this compound was assigned the structure *(94)*. We were surprised that valence-isomerisation to the tetrafluorobenzocyclo-octatetraenamine had apparently not occurred.

(94)

When we allowed pentafluorophenyl-lithium to decompose in ether in the presence of an excess of N,N-dimethylaniline we obtained the compounds *(92)* *(70*, X = F), *(94)*, the latter as the major compound, and a product which was shown to be *(97)*. That this latter compound did not arise by metallation of N,N-dimethylaniline followed by addition to tetrafluorobenzyne was shown by quenching the reaction mixture with deuterium oxide. No deuterium incorporation was detected. The compound *(97)* provides a rare example of a product derived by a Stevens rearrangement in which aryl migration has occurred [b].

(95)   *(96)*   *(97)*

The aryl migration undoubtedly occurs as a result of the strong electron with drawing effect of the four fluorine atoms.

When we carried out a reaction in which we allowed a solution of pentafluorophenyl-lithium in light-petroleum to decompose in the presence of N,N-dimethylaniline we obtained a fifth product. This compound was an extremely weak base and was shown to be *(98)*. Presumably the

---

[b] A number of recent studies [134a–e] have shown that caged radical ions are involved in Stevens rearrangements. Thus a number of systems have been shown to exhibit chemically induced dynamic nuclear polarisation effects in the n.m.r. spectra of reaction mixtures.

change to a non-polar solvent system is responsible for the failure to form the ylide *(96)* and the compound *(98)* arises by immediate neutralisation of the charges present in the zwitter-ion *(95)*.

(98)

10. Reactions with Bicyclohept[2.2.1]ene and Bicyclohepta[2.2.1]diene

The isolation of $(2+2)$ π cyclo-adducts in the reactions of benzyne with bicyclohept[2.2.1]ene and bicyclohepta[2.2.1]diene, and the absence of rearranged products led to the suggestion that concerted cyclo-additions were involved [104]. It is now known that a large number of $(2+2)$ π cyclo-additions involving arynes, although stereoselective, are not stereospecific [115-118,65,109]. In addition a concerted thermal $(2+2)$ π cyclo-addition is not in accord with the conservation of orbital symmetry [68]. We, and others have reinvestigated these reactions. The reactions of the tetra-halogenobenzynes with bicyclohept[2.2.1]ene led to the formation of the exo-adducts *(99,* X = F or Cl), while with bicyclohepta[2.2.1]diene two products *(100,* X = F or Cl) and *(101,* X = F or Cl) were obtained [135].

(99)   *(100)*   *(101)*

The products *(101)* are presumably formed in an orbital symmetry allowed $(2+2+2)$ π concerted cyclo-addition. The formation of the compounds *(99)* and *(100)* are more difficult to rationalise unambiguously. Clearly, if an intermediate such as *(102)* had more than a very transient existence, one would expect to isolate products derived from carbonium ion rearrangements. That no rearranged products have been isolated might be used as an argument for the violation of the Woodward-Hoffmann rules. "There are none!" [68].

It is more likely that the dipolar intermediate *(102)* collapses to the product *(99)* before rearrangement can occur and thus gives rise to a product identical to that expected to be derived in a concerted process. The same type of products have been obtained with a number of other arynes [136,137].

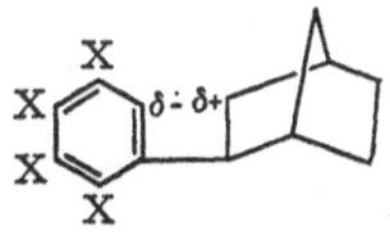

(102)

11. Reactions with Five-Membered Ring Dienes

The reaction of benzyne with furan was the first example of a Diels-Alder reaction of benzyne to be studied [1]. No authenticated examples of arynes are known which fail to give cyclo-adducts with furan [138]. The tetrahalogenobenzynes all form the expected adducts *(103*, X = F [6], Cl [57,139], Br [59] or I [59])*, as do other highly fluorinated arynes [140,141]. The isomeric adducts *(104)* and *(105)* have been detected by [19]F n.m.r. spectroscopy when the dilithio-compound *(12)* was allowed to decompose in the presence of furan [28,103].

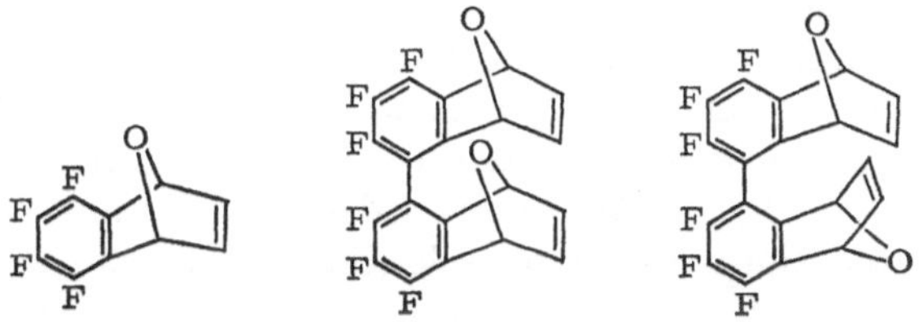

(103)   *(104)*   *(105)*

The reaction of tetrafluorobenzyne with *N*-methylpyrrole leads to a good yield of the adduct *(106)*, and with thiophen to tetrafluoro-naphtha-ene [56]. That this latter reaction was the first example of a Diels-Alder reaction of thiophen was shown by following the reaction by [1]H n.m.r. spectroscopy. Evidence for the intermediacy of the episulphide *(107)* was obtained.

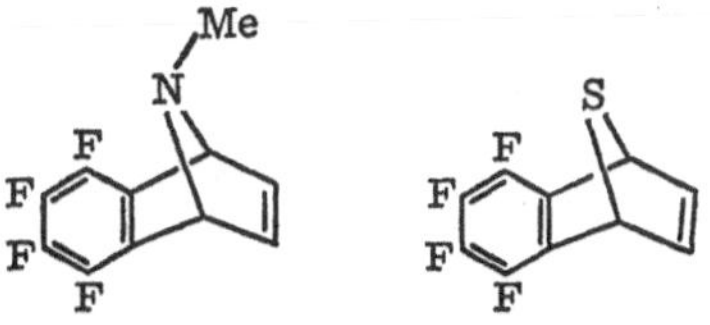

(106)   *(107)*

Evidence for slow inversion at nitrogen in the compound *(106)* and its tetrachloro analogue has been obtained [142]. The adduct of tetrafluorobenzyne with cyclopentadiene has been obtained but of rather more interest is the isolation of the isomeric adducts *(108)* and *(109)* from the reaction of tetrafluorobenzyne with nickelocene [143]. No adducts were obtained from attempted reactions with ferrocene [144,80].

(108)　　　　　　*(109)*

12. Reactions with Steroidal Dienes and Trienes

The reaction of benzyne with cyclohexadiene has been known for some time [1], but although a number of steroidal cis-dienes are readily available no reactions with arynes had been reported prior to our beginning such investigations [145]. This was somewhat surprising in view of the number of reports concerning the modification of steroids by means of reactions with carbenes [146-149] and the known Diels-Alder reactions of steroidal dienes and trienes [150,151].

Benzyne reacts with 7-dehydrocholesteryl methyl ether to form the 'ene' products *(110)* and *(111, X = H)*, while with tetrafluorobenzyne we isolated the anticipated cyclo-adduct *(112)* in addition to an 'ene' product *(111, X = F)* [152]. We expected that less crowding would obtain in the transition states leading to cyclo-adducts from cholesta-2,4-diene, and obtained from a reaction with benzyne both the α- and β-adducts *(113, X = H)* and *(114)*. In a reaction with tetrafluorobenzyne we only obtained the α-adduct *(113, X = F)*.

The formation of the adduct *(112)* undoubtedly reflects the higher energy of tetrafluorobenzyne, as compared with benzyne, and may also possibly result, in part, from a lower transition state energy for the formation of a cyclo-adduct with tetrafluorobenzyne.

(110)　　　　　　*(111)*

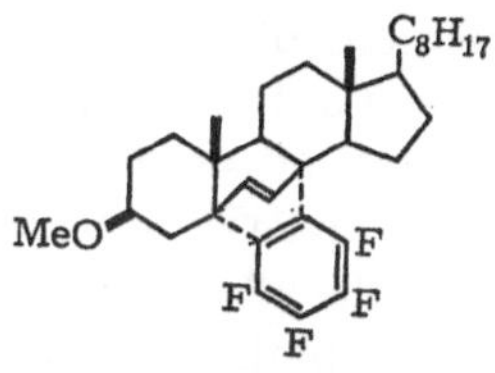

(112) 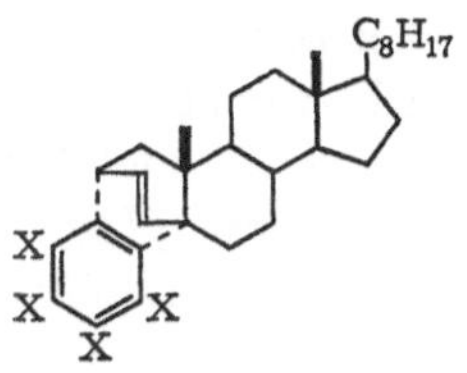   *(113)*

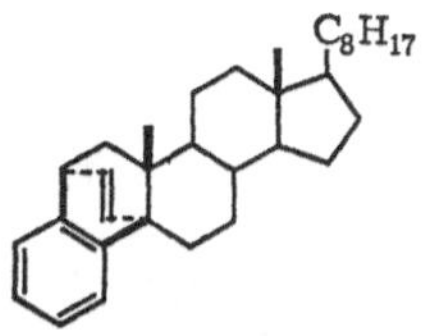

(114)

It had been reported that, for example, 5.7.9(11)-cholestatrienyl acetate shows a greater reactivity towards dienophiles than the corresponding 5,7-dienes [153]. We therefore investigated reactions with tetrafluoro- and tetrachloro-benzyne [152]. The two isomeric adducts *(115)* and *(116)* were obtained in only modest yields with tetrafluorobenzyne, while with tetrachlorobenzyne no adducts or 'ene' products were isolated.

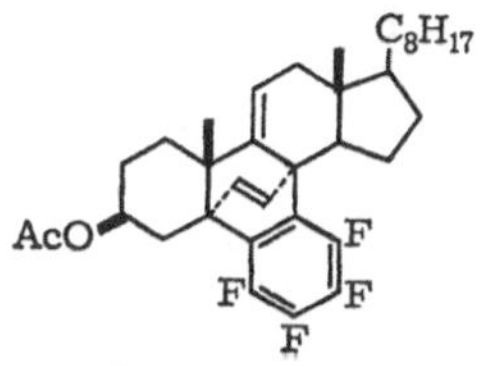

(115) 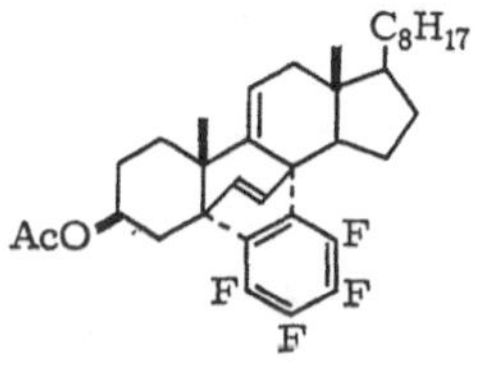   *(116)*

We have also investigated the reactions of benzyne and tetrachlorobenzyne with the rings A/C cis-diene *(117)*, both of which gave the α-adducts *(118,* X = H or Cl) [154].

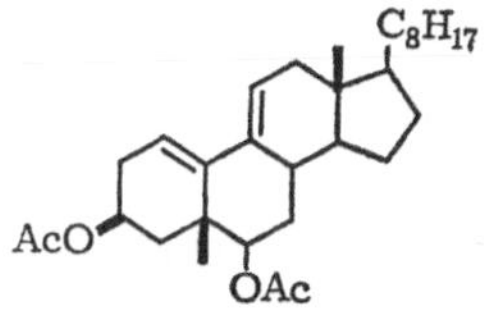

(117) 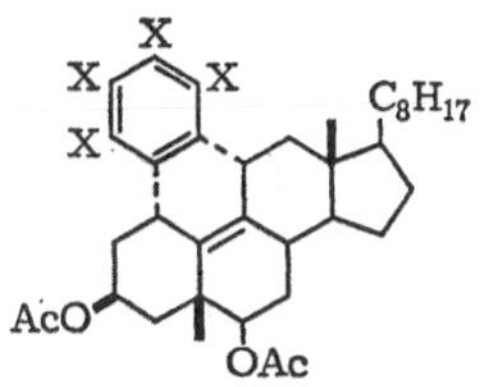   *(118)*

65

In view of our results obtained in reactions of the tetrahalogeno-benzynes with aromatic compounds we carried out a reaction of tetra-fluorobenzyne with the A-ring aromatic steroid 3,17,β-dimethoxy-oestra-1(10),2,4-triene *(119)* [154]. As we expected the initially formed enol-ethers were very rapidly hydrolysed and the adducts were isolated as the ketones *(120)* and *(121)*. The mass spectra of the compounds *(120)* and *(121)* did not show molecular ion peaks and as anticipated the ketones were rapidly converted into the naphthalenes *(122)* and *(123)* photo-chemically.

13. Reactions with Carbonyl Compounds

The availability of the tetrahalogenoanthranilic acids, in particular tetrachloroanthranilic acid, suggested that an investigation of reactions with carbonyl compounds would be worthwhile. The reactions of arynes with for example cinnamaldehyde could in principle give rise to adducts involving the arene residue, the styrene moiety, the α,β-unsaturated aldehyde function, or just one of the reactive groups.

The reaction of tetrachlorobenzyne with cinnamaldehyde gave 5,6,7,8-tetrachloroflavene (124, X = Cl) [155]. Similarly reactions with aliphatic α,β-unsaturated aldehydes gave 2 H-chromen derivatives (125) [156,157].

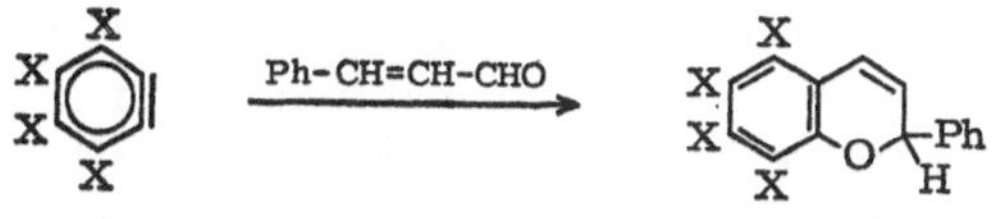

(124)

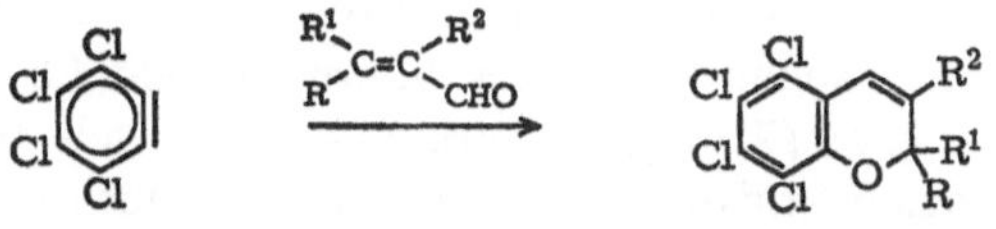

(125)

Evidently, unless migrations have occurred, the isolated products do not arise by 1,4-cyclo-addition reactions. Such migrations are unlikely since they would involve a hydrogen or alkyl group leading to the compounds (125), or an aryl group leading to the compound (124). We suggested [156] that the following mechanism was operating. The formation of an unstable benzo-oxetene (126) by 1,2-cyclo-addition to the carbonyl group would lead to a quinone methide (127) which could then undergo an electrocyclic ring closure to give the isolated products. This type of electrocyclic ring closure has been suggested previously as a possible step in the biosynthesis of the 2,2-dimethylchromen system [158].

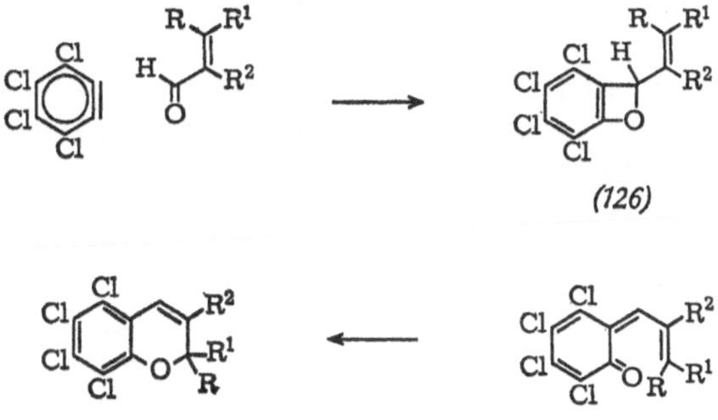

(126)

(125)

(127)

Similar cyclisations have also been invoked recently [159,160]. Support for our suggested mechanism is apparent from the structures of the products obtained with various methyl derivatives of acrolein. Similarly, β-deuteriocinnamaldehyde leads to 2-[^{2}H]5,6,7,8-tetrachloroflavene [157].

It is known that benzenediazonium-2-carboxylate decomposes to give benzyne via the zwitter-ion *(128)* [161]. We therefore checked that benzyne is involved in our reactions by carrying out reactions with cinnamaldehyde using benzyne generated from benzothiadiazole-1,1-dioxide [162], diphenyliodonium-2-carboxylate [163,164], as well as from anthranilic acid [165]. Flavene was isolated from each reaction and hence our reactions do involve arynes and are not arynoid [138].

In the reactions which we have just discussed, the initial reaction involves only the carbonyl group. We therefore looked next at reactions of tetrachlorobenzyne with simpler aldehydes. It is noteworthy that *o*-carboxybenzenediazonium salts are known to yield 1,3-benzodioxan-4-ones, for example, *(129)* with carbonyl compounds [166], and 2,2-diphenyl-3,1-benzo-oxathian-4-one *(130)* with thiobenzophenone [167]. These products are presumably formed *via* reactions with the intermediate *(128)*.

(128)   *(129)*   *(130)*

The aprotic diazotisation of tetrachloroanthranilic acid in the presence of benzaldehyde or *p*-methoxybenzaldehyde results in the formation of the 1,3-benzodioxan derivatives *(131, R = Ph)* and *(131, R = p-C_6H_4-OMe)* respectively [155]. The absence of products analogous to *(129)* in these reactions suggests that the formation of the compounds *(131)* do involve tetrachlorobenzyne. Some indication of the mechanism of these reactions was given by the fact that no analogous adduct has been isolated in attempted reactions of tetrachlorobenzyne with *p*-nitrobenzaldehyde. However, in the presence of acetone we obtained a low yield of the compound *(132)*.

(131)   *(132)*   *(133)*

This result suggests that the 1,4-dipolar intermediate *(133)* is involved in this last reaction, and in accord with this is the formation of a 1:1:1 adduct of tetrachlorobenzyne, acetone and butan-2,3-dione, which has been shown to have the structure *(134)* [168].

(134)

14. Conclusion

We hope that the results outlined in this review have demonstrated that the chemistry of the tetrahalogenobenzynes is sufficiently different from the chemistry of benzyne to be worthy of study. That four electron with drawing substituents are essential to the high reactivity of arynes in reactions with aromatic systems has been demonstrated by generating the isomeric trifluorobenzynes from the aryl-lithium compounds *(135)* and *(136)* in the presence of an excess of benzene [169]. Whereas tetra-fluorobenzyne reacts under similar conditions to give the 1,4-cyclo-adduct in greater than 50% yield the compounds *(137)* and *(138)* are produced in 16% and *ca.* 1.5% yields respectively.

(135)   *(136)*   *(138)*   *(137)*

Similar results have been obtained with other systems, where arynes are known to be formed in good yield [170–172].

I should like to acknowledge the stimulating discussions of problems of mutual interest which I have had with friends and colleagues in the U.S.A. and Germany, as well as in the United Kingdom. I am particularly indebted to Professor G. W. Kirby and Drs. B. A. Marples and K. G. Mason of my own department. This review could not have been written without the patient hard work of enthusiastic co-workers whose names appear in the list of references, and to them I also extend my thanks. I also thank those industrial companies both in England and Germany who have generously donated chemicals; in particular The Imperial Smelting Corporation, Bristol. I also thank Perkin-Elmer Ltd., Beaconsfield, who obtained the ¹H n.m.r. spectra which are repro-duced here.

15. References

1) Hoffmann, R. W.: Dehydrobenzene and Cycloalkynes. Weinheim: Verlag Chemie, G.m.b.H. New York: Academic Press 1967.
2) Pummer, W. J., Wall, L. A.: J. Res. Natl. Bur. Std. 63A, 167 (1959).
3) Nield, E., Stephens, R., Tatlow, J. C.: J. Chem. Soc. 166 (1959).
4) Brooke, G. M., Chambers, R. D., Heyes, J., Musgrave, W. K. R.: J. Chem. Soc. 729 (1964).
5) Pearson, D. E., Cowan, D., Beckler, J. D.: J. Org. Chem. 24, 504 (1959).
6) Coe, P. L., Stephens, R., Tatlow, J. C.: J. Chem. Soc. 3227 (1962).
7) Tamborski, C., Soloski, E. J., Dills, C. E.: Chem. Ind. (London) 2067 (1965).
8) Rausch, M. D., Tibbetts, F. E., Gordon, H. B.: J. Organometal. Chem. (Amsterdam) 5, 493 (1966).
9) Heaney, H., Jablonski, J. M.: Tetrahedron Letters 4529 (1966).
10) Zieger, H. E., Wittig, G.: J. Org. Chem. 27, 3270 (1962).
11) Patrick, C. R., Prosser, G. S.: Nature 187, 1021 (1960).
12) Burdon, J.: Tetrahedron 21, 1101 (1965).
13) Gilman, H., Gorsich, R. D.: J. Am. Chem. Soc. 78, 2217 (1956).
14) Heaney, H., Mann, F. G., Millar, I. T.: J. Chem. Soc. 3930 (1957).
15) — Lees, P.: Tetrahedron Letters 3049 (1964).
16) Bartle, K. D., Heaney, H., Jones, D. W., Lees, P.: Tetrahedron 21, 3289 (1965).
17) Gething, B., Patrick, C. R., Tatlow, J. C.: J. Chem. Soc. 1574 (1961).
18) Belf, L. J., Buxton, M. W., Tilney-Bassett, J. F.: Tetrahedron 23, 4719 (1967).
19) Orndorff, W. R., Nichols, E. H.: Am. Chem. J. 48, 473 (1912).
20) Lesser, R., Weiss, R.: Ber. 46, 3937 (1913).
21) Br. Pat. 860, 292. Chapman, J. H., Graham, W., Evans, R. M.: to Glaxo Laboratories Ltd.
22) Fenton, D. E., Park, A. J., Massey, A. G.: J. Organometal. Chem. (Amsterdam) 2, 437 (1964).
23) — Massey, A. G.: Tetrahedron 21, 3009 (1965).
24) Callander, D. D., Coe, P. L., Tatlow, J. C.: Tetrahedron 22, 419 (1966).
25) Cohen, S. C., Fenton, D. E., Shaw, D., Massey, A. G.: J. Organometal. Chem. (Amsterdam) 8, 1 (1967).
26) — Tomlinson, A. J., Wiles, M. R., Massey, A. G.: J. Organometal. Chem. (Amsterdam) 11, 385 (1968).
27) — — — — Chem. Ind. (London) 877 (1967).
28) — Moore, D., Price, R., Massey, A. G.: J. Organometal. Chem. (Amsterdam) 12, P 37 (1968).
29) Brown, R. F. C., Solly, R. K.: Chem. Ind. (London) 181 (1965); Australian J. Chem. 19, 1045 (1966).
30) Fields, E. K., Meyerson, S.: Chem. Commun., 474 (1965).
31) Cava, M. P., Mitchell, M. J., De Jongh, D. C., Van Fossen, R. Y.: Tetrahedron Letters 2947 (1966).
32) Brown, R. F. C., Gardner, D. V., McOmie, J. F. W., Solly, R. K.: Chem. Commun. 407 (1966).
33) — — — — Australian J. Chem. 20, 139 (1967).
34) Gardner, D. V., McOmie, J. F. W., Albriktsen, P., Harris, R. K.: J. Chem. Soc. (C) 1994 (1969).
35) Sartori, P., Golloch, A.: Chem. Ber. 102, 1763 (1969).
36) — Weidenbruch, M.: Angew. Chem. 77, 1076 (1965); Angew. Chem. Intern. Ed. Engl. 4, 1072 (1965).
37) — — Chem. Ber. 100, 3016 (1967).

38) Miller, R. G., Stiles, M.: J. Am. Chem. Soc. *85*, 1798 (1963).
39) Friedman, L.: J. Am. Chem. Soc. *89*, 3071 (1967).
40) Hantzsch, A., Davidson, W. B.: Ber. *29*, 1535 (1896).
41) Stiles, M., Miller, R. G., Burckhardt, U.: J. Am. Chem. Soc. *85*, 1792 (1963).
42) Campbell, C. D., Rees, C. W.: J. Chem. Soc. (C) 742 (1969).
43) Vorozhtsov, N. N., Barkhasch, V. A., Ivanova, N. G., Petrov, A. K.: Tetrahedron Letters 3575 (1964).
44) Brewer, J. P. N., Heaney, H.: Tetrahedron Letters 4709 (1965).
45) Filler, R.: Illinois Institute of Technology, Chicago, U.S.A., Personal communication.
46) Vorozhtsov, I. N., Ivanova, N. G., Barkhash, V. A.: Zh. Organ. Khim. *3*, 220 (1967).
47) Callander, D. D., Coe, P. L., Tatlow, J. C.: Chem. Commun. 143 (1966).
48) Cohen, S. C., Fenton, D. E., Tomlinson, A. J., Massey, A. G.: J. Organometal. Chem. (Amsterdam) *6*, 301 (1966).
49) Petrova, T. D., Savchenko, T. I., Yakobson, G. G.: Zh. Obshch. Khim. *37*, 1170 (1967); J. Gen. Chem. USSR *37*, 1110 (1967).
50) Brewer, J. P. N., Heaney, H.: Chem. Commun. 811 (1967).
51) See for example Zimmerman, H. E., Givens, R. S., Pagni, R. M.: J. Am. Chem. Soc. *90*, 6096 (1968).
52) Hahn, R. C., Rothman, L. J.: J. Am. Chem. Soc. *91*, 2409 (1969); and references cited.
53) Eckhard, I. F., Heaney, H., Marples, B. A.: Tetrahedron Letters 3273 (1969).
54) Brewer, J. P. N., Eckhard, I. F., Heaney, H., Marples, B. A.: J. Chem. Soc. (C) 664 (1968).
55) Vorozhtsov, I. N., Ivanova, N. G., Barkhash, V. A.: Izv. Akad. Nauk SSSR, Ser. Khim. 1514 (1967).
56) Callander, D. D., Coe, P. L., Tatlow, J. C., Uff, A. J.: Tetrahedron *25*, 25 (1969).
57) Heaney, H., Jablonski, J. M.: J. Chem. Soc. (C) 1895 (1968).
58) Berry, D. J., Wakefield, B. J.: J. Chem. Soc. (C) 2342 (1969).
59) Heaney, H., Mason, K. G., Sketchley, J. M.: unpublished observations.
60) Holmes, J. M., Peacock, R. D., Tatlow, J. C.: Proc. Chem. Soc. (London) 108 (1963).
61) Massey, A. G., Randall, E. W., Shaw, D.: Chem. Ind. (London) 1244 (1963).
62) Chambers, R. D., Chivers, T.: Organometal. Chem. Rev. *1*, 279 (1966).
63) Tomlinson, A. J., Massey, A. G.: J. Organometal. Chem. (Amsterdam) *8*, 321 (1967).
64) Atkin, R. W., Rees, C. W.: Chem. Commun. 152 (1969).
65) Jones, M., Levin, R. H.: J. Am. Chem. Soc. *91*, 6411 (1969).
66) Hoffmann, H. M. R., Angew. Chem. *81*, 597 (1969); Angew. Chem. Intern. Ed. Engl. *8*, 556 (1969).
67) Hoffmann, R., Imamura, A., Hehre, W. J.: J. Am. Chem. Soc. *90*, 1499 (1968).
68) Woodward, R. B., Hoffmann, R.: Angew. Chem. *81*, 888 (1969); Angew. Chem. Internat. Ed. Engl. *8*, 781 (1969).
69) Bryce-Smith. D.: Chem. Commun. 806 (1969).
70) Kampmeier, J. A., Hoffmeister, E.: J. Am. Chem. Soc. *84*, 3787 (1962).
71) Kharasch, N., Sharma, R. K.: Chem. Commun. 492 (1967).
72) Sharma, R. K., Kharasch, N.: Angew. Chem. *80*, 69 (1968); Angew. Chem. Intern. Ed. Engl. *7*, 36 (1968).
73) Heaney, H., Johnson, M. G., Ward, T. J.: unpublished observations.
74) Wilzbach, K. E., Kaplan, L.: J. Am. Chem. Soc. *88*, 2066 (1966).
75) Bryce-Smith, D., Gilbert, A., Orger, B. H.: Chem. Commun. 512 (1966).

76) Brewer, J. P. N., Heaney, H., Marples, B. A.: Chem. Commun. 27 (1967).
77) Jackman, L. M., Sternhell, S.: Applications of Nuclear Magnetic Resonance Spectroscopy in Organic Chemistry. London: Pergamon 1969.
78) Sternhell, S.: Rev. Pure Appl. Chem. *14*, 15 (1964).
79) Pople, J. A., Schneider, W. G., Bernstein, H. J.: High-resolution Nuclear Magnetic Resonance New York: McGraw-Hill 1959.
80) Brewer, J. P. N., Heaney, H.: unpublished observations.
81) Heaney, H., Mason, K. G., Sketchley, J. M., Ward, T. J.: unpublished observations.
82) Cook, J. D., Wakefield, B. J., Clayton, C. J.: Chem. Commun. 150 (1967).
83) — — Tetrahedron Letters 2535 (1967).
84) — — Heaney, H., Jablonski, J. M.: J. Chem. Soc. (C) 2727 (1968).
85) Pauling, L.: The Nature of the Chemical Bond, 3rd Edn. Oxford 1960.
86) Grützmacher, H.-F., Hübner, J.: Org. Mass Spectr. *2*, 649 (1969).
87) Klandermann, B.: J. Am. Chem. Soc. *87*, 4649 (1965).
88) — Criswell, T. R.: J. Am. Chem. Soc. *91*, 510 (1969).
89) — — J. Org. Chem. *34*, 3426 (1969).
90) Hankinson, B., Heaney, H.: unpublished observations.
91) Baker, W., Barton, J. W., McOmie, J. F. W.: unpublished experiments reported in D. Ginsburg (Ed.), Non-Benzenoid Aromatic Compounds p. 78. New York: Intersience 1959.
92) Farnum, D. G., Atkinson, E. R., Lothrop, W. C.: J. Org. Chem. *26*, 3204 (1961).
93) Blatchly, J. M., McOmie, J. F. W.: personal communication.
94) Barton, J. W.: personal communication.
95) Roberts, J. D.: Molecular Orbital Calculations, p. 103. New York: Benjamin 1962.
96) Heaney, H., Mason, K. G., Sketchley, J. M.: Tetrahedron Letters 485 (1970).
97) Brewer, J. P. N., Heaney, H., Marples, B. A.: Tetrahedron *25*, 243 (1969).
98) Davies, W., Wilmshurst, J. R.: J. Chem. Soc. 4079 (1961).
99) Corbett, T. G., Porter, Q. N.: Australian J. Chem. *18*, 1781 (1965).
100) Dyke, S. F., Marshall, A. R., Watson, J. P.: Tetrahedron *22*, 2515 (1966).
101) Harrison, R., Heaney, H., Jablonski, J. M., Mason, K. G., Sketchley, J. M.: J. Chem. Soc. (C) 1684 (1969).
102) Povolotskaya, N. N., Limasova, T. I., Vorozhtsov, I. N., Barkhash, V. A.: Zh. Obshch. Khim. *38*, 1651 (1968).
103) Cohen, S. C., Reddy, M. L. N., Roe, D. M., Tomlinson, A. J., Massey, A. G.: J. Organometal. Chem. (Amsterdam) *14*, 241 (1968).
104) Simmons, H. E.: J. Am. Chem. Soc. *83*, 1657 (1961).
105) Dilling, W. L.: Tetrahedron Letters 939 (1966).
106) Ciganek, E.: J. Org. Chem. *34*, 1923 (1969).
107) Cram, D. J., Willey, F., Fischer, H. P., Relles, H. M., Scott, D. A.: J. Am. Chem. Soc. *88*, 2759 (1966).
108) — Uyeda, R. T.: J. Am. Chem. Soc. *84*, 5466 (1964), and references cited therein.
109) Jones, M., Levin, R. H.: Tetrahedron Letters 5593 (1968).
110) Norman, R. O. C., Taylor, R.: Electrophilic Substitution in Benzenoid Compounds, p. 301. Amsterdam: Elsevier 1965.
111) Gould, E. S.: Mechanism and Structure in Organic Chemistry, p. 436. New York: Holt, Rinehart and Winston 1959.
112) March, J.: Advanced Organic Chemistry, p. 387. New York: McGraw-Hill 1968.
113) Bryce-Smith, D., Perkins, N. A.: J. Chem. Soc. 5295 (1962).
114) Hankinson, B., Heaney, H.: Tetrahedron Letters 1335 (1970).

115) Wasserman, H. H., Solodar, J.: J. Am. Chem. Soc. *87*, 4002 (1965).
116) Tabushi, I., Oda, R., Okazaki, K.: Tetrahedron Letters 3743 (1968).
117) Wasserman, H. H., Solodar, A. J., Keller, L. S.: Tetrahedron Letters 5597 (1968).
118) Friedman, L., Osiewicz, R. J., Rabideau, P. W.: Tetrahedron Letters 5735 (1968).
119) Murray, R. K., Hart, H.: Tetrahedron Letters 4995 (1968).
120) Hart, H., Murray, R. K.: Tetrahedron Letters 379 (1969).
121) Ipaktschi, J.: Tetrahedron Letters 215 (1969).
122) Givens, R. S., Oettle, W. F.: Chem. Commun. 1164 (1969).
123) Font, J., Serratosa, F., Vilarrasa, L.: Tetrahedron Letters 4743 (1969).
124) Franzen, V., Joschek, H.-I., Mertz, C.: Annalen *654*, 82 (1962).
125) Hellmann, H., Eberle, D.: Annalen *662*, 188 (1963).
126) Brewer, J. P. N., Heaney, H., Ward, T. J.: J. Chem. Soc. (C) 355 (1969).
127) Chambers, R. D., Spring, D. J.: Tetrahedron Letters 2481 (1969).
128) Brewer, J. P. N., Heaney, H., Jablonski, J. M.: Tetrahedron Letters 4455 (1968).
129) Baldwin, J. E., Hackler, R. E., Kelly, D. P.: Chem. Commun, 538 (1968).
130) Logullo, F. M.: Ph. D. Thesis, Case Institute of Technology, quoted in reference 1, p. 178.
131) Wolthuis, E., Bouma, B., Modderman, J., Sytsma, L.: Tetrahedron Letters 407 (1970).
132) Lepley, A. R.: Preprints, American Chemical Society, Division of Petroleum Chemistry, 1969, *14* (No. 2), April 1969, C 43.
133) — Giumanini, A. G., Giumanini, A. B., Khan, W. A.: J. Org. Chem. *31*, 2051 (1966).
134) Heaney, H., Ward, T. J.: Chem. Commun. 810 (1969).
134) a) Jemison, R. W., Morris, D. G.: Chem. Commun. 1126 (1969);
   b) Lepley, A. R.: J. Am. Chem. Soc. *91*, 1237 (1969).
   c) — Chem. Commun. 1460 (1969;)
   d) Schöllkopf, W., Ostermann, G., Schossig, J.: Tetrahedron Letters 2619 (1969);
   e) — Ludwig, U., Ostermann, G., Patsch, M.: Tetrahedron Letters 3515 (1969).
135) Heaney, H., Jablonski, J. M.: Tetrahedron Letters 2733 (1967).
136) Friedman, L.: personal communication.
137) Baker, R., Mason, T. J.: Chem. Commun. 120 (1969).
138) See for example Cadogan, J. I. G., Cook, J., Harger, M. J. P., Sharp, J. T.: Chem. Commun. 299 (1970); and references cited therein.
139) Rausch, M. D.: personal communication.
140) Tamborski, C., Soloski, E. J.: J. Organometal. Chem. (Amsterdam) *10*, 385 (1967).
141) De Pasquale, R. J., Tamborski, C.: J. Organometal. Chem. (Amsterdam) *13*, 273 (1968).
142) Gribble, G. W., Easton, N. R., Eaton, J. T.: Tetrahedron Letters 1075 (1970).
143) Roe, D. M., Massey, A. G.: J. Organometal. Chem. (Amsterdam) *20*, Pl (1969).
144) Massey, A. G.: personal communication.
145) Eckhard, I. F., Heaney, H., Marples, B. A.: Tetrahedron Letters 4001 (1967).
146) Knox, L. H., Valarde, E., Berger, S., Cuadriello, D., Landis, P. W., Cross, A. D.: J. Am. Chem. Soc. *85*, 1851 (1963).
147) Birch, A. J., Subba Rao, G. S.: Tetrahedron *7*, 391 (1966).
148) Nazer, M. S.: J. Org. Chem. *30*, 1737 (1965).
149) Bond, F. T., Cornelia, R. H.: Chem. Commun. 1189 (1968).
150) Fieser, L. F., Fieser, M.: Steroids, p. 109, 265. New York: Reinhold 1959.

[151] Lakeman, J., Speckamp, W. N., Huisman, H. O.: Tetrahedron *24*, 5151 (1968), and references cited therein.

[152] Eckhard, I. F., Heaney, H., Marples, B. A.: J. Chem. Soc. (C) 2098 (1969).

[153] van der Gen, A., Zunnebeld, W. A., Pandit, U. K., Huisman, H. O.: Tetrahedron *21*, 3651 (1965).

[154] Eckhard, I. F., Heaney, H., Marples, B. A.: J. Chem. Soc. (C), in press (1970).

[155] Heaney, H., McCarty, C. T.: Chem. Commun. 123 (1970).

[156] — Jablonski, J. M.: Chem. Commun. 1139 (1968).

[157] — McCarty, C. T.: unpublished observations.

[158] Ollis, W. D., Sutherland, I. O.: Chemistry of Natural Phenolic Compounds, p. 84. Oxford: Pergamon Press 1961.

[159] Schweizer, E. E., Crouse, D. M., Dalrymple, D. L.: Chem. Commun. 354 (1969), and references cited therein.

[160] Hug, R., Hauser, H.-J., Schmid, H.: Chimia (Aarau) *23*, 108 (1969).

[161] Gompper, R., Seybold, G., Schmolke, B.: Angew. Chem. *80*, 404 (1968); Angew. Chem. Intern. Ed. Engl. *7*, 389 (1968).

[162] Wittig, G., Hoffmann, R. W.: Chem. Ber. *95*, 2718 (1962).

[163] Le Goff, E.: J. Am. Chem. Soc. *84*, 3786 (1962).

[164] Berringer, F. M., Huang, S. J.: J. Org. Chem. *29*, 445 (1964).

[165] Friedman, L., Logullo, F. M.: J. Am. Chem. Soc. *85*, 1549 (1963).

[166] Chiusoli, G. P., Venturello, C.: Chem. Commun. 771 (1969).

[167] Dittmer, D. C., Whitman, E. S.: J. Org. Chem. *34*, 2004 (1969).

[168] Heaney, H., McCarty, C. T.: Chem. Commun. 123 (1970).

[169] Harrison, R., Heaney, H.: J. Chem. Soc. (C) 889 (1968).

[170] Fernandez, R. A., Heaney, H., Jablonski, J. M., Mason, K. G., Ward, T. J.: J. Chem. Soc. (C) 1908 (1969).

[171] Cook, J. D., Wakefield, B. J.: J. Chem. Soc. (C) 1973 (1969).

[172] Berry, D. J., Collins, I., Roberts, S. M., Suschitzky, H., Wakefield, B. J.: J. Chem. Soc. (C) 1285 (1969).

Received June 9, 1970

Hetero-Cope-Reaktionen

Prof. Dr. E. Winterfeldt

Organisch-Chemisches Institut der Universität Hannover

Inhalt

I. Einleitung

Nachdem die ersten Beispiele der nach dem allgemeinen Schema *1* ablaufenden „Cope-Umlagerung" bekannt geworden waren [1], wurde der Mechanismus dieser Reaktion unter verschiedenen Aspekten untersucht und auf verschiedene Weisen gedeutet [2-4].

G. S. Hammond [3] hat in dieser Situation Sehergaben bewiesen, als er dazu aufforderte, doch zunächst alle Reaktionen, die nach *1* verlaufen, unabhängig vom Mechanismus als „Cope-Umlagerung" zu bezeichnen.

$$A \overset{B}{\underset{\underset{B'}{\overset{|}{A'}}}{\diagdown}} C \qquad \longrightarrow \qquad A \overset{B}{\underset{\underset{B'}{A'}}{\diagup}} C$$

1

Es konnten nämlich, als bald darauf R. B. Woodward und R. Hoffmann ihr Konzept von der Erhaltung der Orbital-Symmetrie entwickelten [5-9], Umwandlungen dieses Typs zwanglos in die Gruppe der Sigmatropen Reaktionen eingegliedert werden. Man versteht unter der Bezeichnung „Sigmatrope Reaktion der Ordnung [i, J]" definitionsgemäß die als nicht katalysierten, intramolekularen Prozeß ablaufende Wanderung einer σ-Bindung, die von einem oder mehreren π-Elektronensystemen flankiert ist, in eine neue Position, in der die durch die neue Bindung verknüpften Zentren um i bzw. J. Atome von den ursprünglichen Begrenzungspunkten der Bindung entfernt liegen.

Die Cope-Umlagerung ist demnach eine sigmatrope Reaktion der Ordnung [3,3] und wird verstanden als Bildung und gleichzeitige Rekombination zweier Allylradikale. Da 2 das höchste besetzte Orbital eines Allylradikals repräsentiert, wird 3 den Übergangszustand der symmetrieerlaubten sigmatropen [3,3] Reaktion wiedergeben.

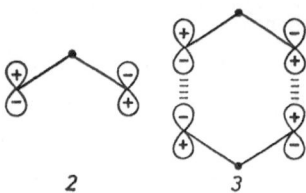

2   3

Unabhängig von den Atomen, die diesen Übergangszustand bilden, wird also ein concertierter thermischer Umlagerungsprozeß zu erwarten sein und neben der „Cope-Reaktion", bei der C-Atome den Übergangszustand bilden, gibt es die große Gruppe der sog. Hetero-Cope-Reaktionen, bei denen ein oder mehrere C-Atome durch Heteroatome ersetzt sind.

Das Ziel dieser Zusammenfassung ist es, einen Überblick über die neuere Entwicklung der 3,3-sigmatropen Reaktionen unter Einbeziehung von Heteroatomen in das umlagerungsfähige System zu geben [10]. Besonders wird dabei das Augenmerk auf eben solche Substrate zu richten sein, bei denen eine, auf diese Weise herbeigeführte, Verminderung der Bindungsenergie für die zu lösende σ-Bindung zu einer Absenkung der Aktivierungsschwelle im Übergangszustand führt, so daß die entsprechenden Umlagerungen bereits bei relativ niedriger Temperatur ablaufen.

II. Claisen-Analoga

Nach der benutzten Nomenklatur gehören auch so etablierte und gut studierte Reaktionen wie die Claisen-Allyläther-Umlagerung 4 und die thermische Fischer-Indolsynthese 5 zur Gruppe der Hetero-Cope-Reaktionen. Da beide Reaktionen vor kurzem in kompetenter Weise in Übersichtsartikeln behandelt worden sind [11,12,14a)] sollen sie hier nicht im

Detail abgehandelt werden, sondern es bleibt nur noch die Aufgabe, einige neuere Entwicklungen hinzuzufügen.

4

5

So wurde kürzlich ein Beispiel einer Claisenanalogen-Umlagerung mitgeteilt, bei der durch Verwendung des Benzyl-vinyläthers *6* eine Umkehrung der üblichen Verhältnisse herbeigeführt wird und der Aromat die Allyläther-Komponente liefert [13].

6   *7*   *8*   *9*

Über *7* bildet sich thermisch das Keton *8*. Der Benzyl-phenoläther *9* erweist sich jedoch als stabil, da offenbar die gleichzeitige Störung zweier aromatischer Systeme nicht erzwungen wird.

Neben vielen wichtigen Arbeiten der Gruppe um H. Schmid in Zürich zur Stereochemie dieser Reaktion [11,14,14a] sei die präparativ interessante und auf einer Sequenz von Cope- und Claisen-Umlagerung basierende Technik von Thomas erwähnt, die bereits bei der Synthese des Sinensals so erfolgreich angewendet werden konnte [15], und deren Übertragung auf Furanderivate kürzlich, ausgehend vom Vinyläther *10*, über die Zwischenstufen *11—16* die Synthese des Torreyals *17* ermöglichte [16].

Ebenfalls als Sequenz mehrerer sigmatroper Prozesse bietet sich die Spaltung ungesättigter Acetale dar, über die eine neuere detaillierte Untersuchung französischer Autoren vorliegt [17]. Das allgemeine Verhalten ist geprägt durch die beiden wichtigen Umwandlungen *18 → 19* und *20 → 21*.

So liefert *22* den Vinyläther *23*, der in einem Folgeprozeß zu *24* und *25* führt.

Ebenfalls über zwei Umlagerungsschritte geht *26* über *27* in *28* über.

Auch das synthetisch reizvolle Verfahren von W. S. Johnson und seiner Gruppe muß erwähnt werden, das jüngst unter Verwendung von Orthoestern zu den umlagerungsfähigen Zwischenstufen *29* führte und offenbar trans-trisubstituierten Olefine *30* mit hoher Stereoselektivität zugänglich macht [18].

Nach Umwandlung des Esters *30* in einen Aldehyd (*31*) kann durch Umsetzung mit 2-Propenyllithium der zur Erzeugung der nächsten Doppelbindung notwendige Allylalkohol erhalten werden.

Interessante präparative Möglichkeiten bietet auch die von H. Pommer u. Mitarb. erarbeitete Erweiterung der sog. Kimel-Cope-Reaktion, deren Ausdehnung auf Acylmalonester über die Ketoester vom Typ *33* wichtige Zwischenprodukte zugänglich macht [19].

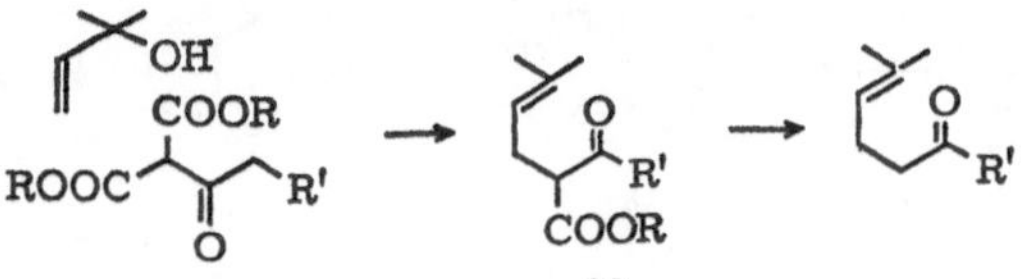

33

Auch weitere Untersuchungen zur Stereochemie der aliphatischen Claisen-Umlagerungen sind inzwischen publiziert worden [14,20,21].

Daß die in ihrer klassischen Form an Phenyl-Allyläthern durchgeführte Reaktion auch auf diverse heterocyclische Aromaten zu übertragen ist, geht schon aus dem bereits zitierten Beispiel von Thomas hervor [16]. Aber die Vielfalt soll doch durch einige weitere Umwandlungen demonstriert werden, die in den letzten Jahren bekannt geworden sind.

So liefert *34* bei 250 °C *35* und unter Einbeziehung des Stickstoffatoms *(36)* zu etwa gleicher Menge *37* [22,23].

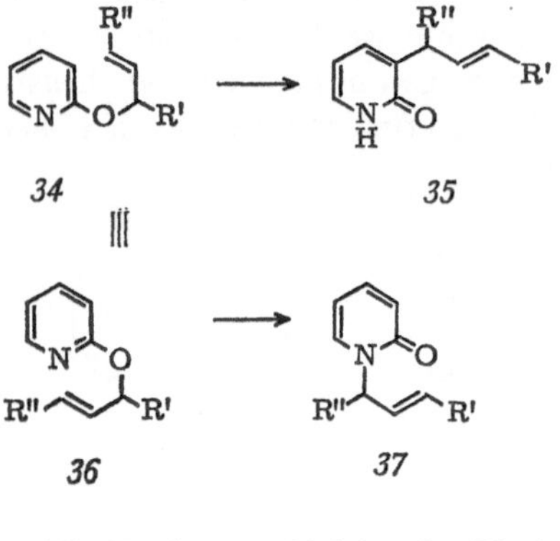

34　　　　*35*

36　　　　*37*

Bei den 4-Oxypyridin-Derivaten *38* folgt der Umlagerung eine Cyclisierung zum Dihydrofuran *40*.

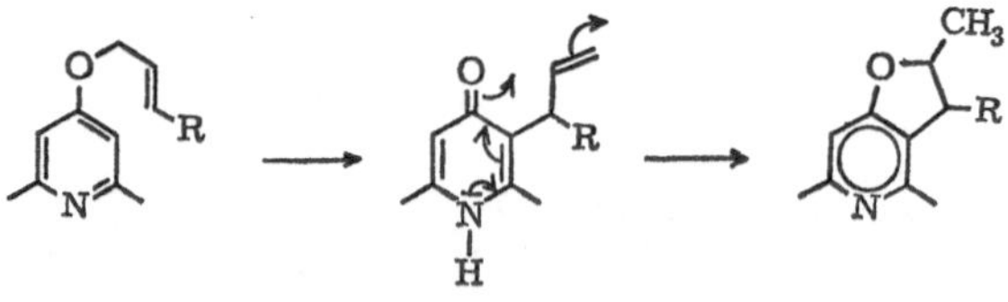

38　　　*39*　　　*40*

Im Gegensatz zu Pyridinderivaten, bei denen benachbartes N- wie C-Atom in gleichem Maße Bestandteil des Umlagerungssystems sein können, findet Makisumi beim Chinolinderivat *41* ausschließlich Umlagerung zur N-Allylverbindung *42* ohne konkurrierende Beteiligung der 3-Position [24]. Dieses Resultat steht in guter Übereinstimmung mit entsprechenden Beobachtungen in der Naphthalinreihe [25].

Win und Tiekelmann bestätigen diesen Befund auch an Isochinolinäthern. Während *43* bei 250 °C in 94% Ausbeute die N-Alkylverbindung *44* erzeugt, geht *45* in die C-Alkylverbindung *46* über [26].

Auf weitere Beispiele, die unter Einbeziehung eines zweiten Heteroatoms verlaufen, wird weiter unten eingegangen werden. Hier soll nur noch ein Hinweis auf das Chromon-System *47* gegeben werden, dessen Umlagerung zu *48* mehrfach untersucht wurde [27–29].

Im übrigen kann dieser Aspekt der Claisen-Umlagerung an heterocyclischen Aromaten mit diesen wenigen Schlaglichtern abgeschlossen werden, denn es existiert hier bereits eine neue Zusammenfassung von Y. Makisumi [30], dessen Gruppe einen erheblichen Beitrag zu diesem Gebiet geleistet hat [31,32].

a) Thio-Claisen-Reaktionen

Es liegt auf der Hand, das Sauerstoffatom der hier besprochenen 3,3-sigmatropen Prozesse durch ein Schwefelatom zu ersetzen. Dieses Experiment ist in der Tat bereits vor einigen Jahren von H. Kwart durchgeführt worden [33,34]. Jedoch resultierten aus *49* die cyclischen Verbindungen *50* und *51*.

Dieser Ausgang läßt mehrere Deutungsmöglichkeiten zu, die Tatsache jedoch, daß es kürzlich gelang, die ringoffene Allylverbindung *52* durch gleichzeitige Einwirkung von Methyljodid und Alkali abzufangen, schafft hier Klärung, und man erkennt, daß die cyclischen Produkte in einer Folgereaktion nach normal ablaufendem sigmatropem Prozeß gebildet werden [35].

Auch bei der Übertragung auf heterocyclische Aromaten erhält man aus *53* die cyclischen Thioäther *55*. In Gegenwart von Säureanhydrid aber wird die Thiol-Zwischenstufe durch Acylierung zu *54* abgefangen und als Primärprodukt erkannt [31,32,36,37].

In beiden Fällen kann durch alkalische Spaltung des Thiolesters die Thiolverbindung regeneriert und ihre glatte Cyclisierung demonstriert werden.

Ebenfalls mit begleitender Cyclisierung verläuft die Thio-Claisen-Umlagerung bei den Thiophenderivaten *59* und *61* [38]).

Die zu erwartenden Allen-Thione stabilisieren sich unter Aromatisierung und Addition, wobei jedoch in diesem Falle ausschließlich das terminale C-Atom angegriffen wird, so daß sich 6-Ring-Äther und nicht etwa Thiopheno-Thiophene bilden*.

Verhindert man durch geeignete Substitution die Ausbildung der aromatischen Thiol-Form, so kann man sowohl Allyl- als auch Allenyl-Umlagerungsprodukte isolieren, wie es beispielsweise Bycroft und Landon in der Indolserie an den Verbindungen *63* und *65* demonstrierten [39]).

Es wurden die direkten Umlagerungsprodukte *64* und *66* isoliert und durch Deuteriummarkierung in *65* der cyclische Prozeß belegt.

Da bei dieser Reaktion im Falle der unsubstituierten Verbindung *67* der sigmatrope Prozeß bereits bei Raumtemperatur durch die NMR-

---

* *Anmerkung bei der Korrektur:* In einer neueren Mitteilung [L. Brandsma u. D. Schuijl-Laros: Recueil *89*, 110 (1970)] wird jedoch über die Bildung der entsprechenden Thienothiophene bei der Behandlung mit Diisopropylamin in Dimethylsulfoxid berichtet.

Technik nachweisbare Mengen an *68* hervorbringt, läßt sich das Primärprodukt direkt nachweisen.

Bei erhöhter Temperatur wird dann nicht etwa die Bildung von *68*
beschleunigt und die Ausbeute erhöht, sondern es läuft dann die als Alternative zu erwartende Radikal-Rekombination zu *69* dem sigmatropen
Prozeß den Rang ab.

Dieses Phänomen der Bevorzugung des nicht-sigmatropen — also die
höhere Aktivierungsenergie verlangenden — Prozeß bei erhöhter Temperatur ist in neuerer Zeit an einer großen Zahl von Umlagerungen von
Baldwin u. Mitarb. demonstriert worden [40–42].

Iminoäther mit blockierter β-Position *(70)* liefern in dieser Serie die
N-Allylverbindung *71*.

Natürlich fehlt es auch nicht an Beispielen aus der acyclischen Reihe.
Es sind hier vor allem die interessanten Reaktionen der Arbeitsgruppe
um Brandsma an Ketenthioacetalen, die gleichzeitig bemerkenswerte
präparative Möglichkeiten offenlegen [43,44].

Die aus den Thioestern vom Typ *72* gut zugänglichen Verbindungen
73 und *74* liefern selbst bei den thermisch stabileren Acetylenverbindun-
gen *74* bereits beim Erhitzen auf 100 bis 120 °C nach 15 min die Umlagerungsprodukte *75* bzw. *77*. Während die basenkatalysierte Cyclisierung
von *77* wiederum am terminalen C-Atom angreifend *76* erzeugt, führt
Säurekatalyse zum Thiophenderivat *78*.

Die durch das Heteroatom ausgelöste Absenkung der Aktivierungsschwelle ist an diesem Beispiel bereits klar erkennbar. Während die
normale Claisen-Umlagerung Aktivierungsberge von etwa 30—35 Kcal/
Mol zu überwinden hat [45,46], zeigt die UV-spektroskopische Verfolgung
dieser Thio-Claisen-Reaktion für den unsubstituierten Fall eine Aktivierungsenergie von 20,7 Kcal/Mol an. Von weiteren Substraten, die ver-
gleichbare maximale Aktivierungsenergien für die Umlagerung verlan-
gen, wird weiter unten die Rede sein.

Noch reaktionsfreudiger sind die entsprechenden Acetylen-thioäther *79*, die man zwar aus dem Natrium-Acetylenthiolat und Allylbromid darstellen kann, die aber bereits beim Eindampfen bei Raumtemperatur exotherm polymerisieren. Die Vermutung, daß das instabile Thioketen *80* als Zwischenstufe durchlaufen wird, ließ sich durch Abfangreaktion mit sekundären Aminen bestätigen. Man isolierte die erwarteten Thioamide *81*, deren Struktur durch spektrale Daten überzeugend belegt wurde [47].

b) AZA-Claisen-Reaktionen

Tauscht man das Sauerstoffatom des Claisen-Systems gegen Stickstoff aus, so erhält man die Verbindung *82*, die in einem thermischen Prozeß das Imin *83* liefert. Eine Substanz, die bei saurer Hydrolyse den Aldehyd *84* liefert.

Dieses Experiment ist kürzlich von Hill und Gilman durchgeführt worden [48].

Mit der substituierten Verbindung *85* gelangt man zu einem Gemisch der cis-trans isomeren Olefine *87* und *88*, in dem die trans-Verbindung *87* mit 87% deutlich überwiegt.

Einmal mehr wird hier die Überlegenheit des Sessel-Übergangszustandes mit equatorialer Methylgruppe *(89)* gegenüber dem mit axialer Methylgruppe *(90)* demonstriert.

Natürlich bietet die Reihe der Heterocyclen auch ein weites Feld für Umlagerungen dieser Art. So liefert *91* bei 180 °C *92* [49] und *93* analog *94* [50].

Gut geeignete Ausgangsmaterialien sind auch die Anhydrobasen *95* und *97*, die mit Alkoholat aus den entsprechenden Quartärsalzen hervorgehend glatt *96* und *98* erzeugen [51].

95   *96·*

97   *98*

Von der Möglichkeit, das aromatische System in den Umlagerungsprozeß einzubeziehen, wird offenbar kein Gebrauch gemacht, obwohl das grundsätzlich möglich wäre. Da bei einem Molekül, bei dem Bruch einer N-O-Bindung den Prozeß auslöst, die Beobachtung gemacht worden war (s. Seite 94), daß ein aromatisches System einer olefinischen Doppelbindung durchaus bei der Beteiligung am Übergangszustand den Rang ablaufen kann, wurden die Enamine *99* und *100* untersucht, bei denen jeweils 2 Reaktionsmöglichkeiten offenstehen [52].

In beiden Fällen wird auch hier ausschließlich das über den Übergangszustand *99a* bzw. *100a* gebildete Produkt isoliert. Man erhält jedoch nicht die Primärprodukte. *102* geht unter den Reaktionsbedingungen sofort in das thermdoynamisch stabilere Fumarester-derivat *101* über, das ebenfalls nur als Nebenprodukt isoliert wird. Hauptprodukt ist das Cyclisierungsprodukt *104*, das in hoher Ausbeute direkt auskristallisiert.

Auf das intermediäre Auftreten der Allenverbindung *103* kann man aus dem isolierten Pyrrolderivat *105* schließen. Die optimale Reaktionstemperatur liegt hier bei 180 °C und es drängt sich auf, nach konstitutionellen Veränderungen Ausschau zu halten, die die Reaktionstemperatur herabsetzen.

Bereits bei 140 °C geht unter gleichzeitiger Öffnung des Aziridin-Systems *106* in *107* über [53] und ebenfalls unter Öffnung eines Cyclopropanringes verläuft die von H. A. Staab u. Mitarb. detailliert untersuchte Umlagerung der Dialdimine vom Typ *108* [54,55].

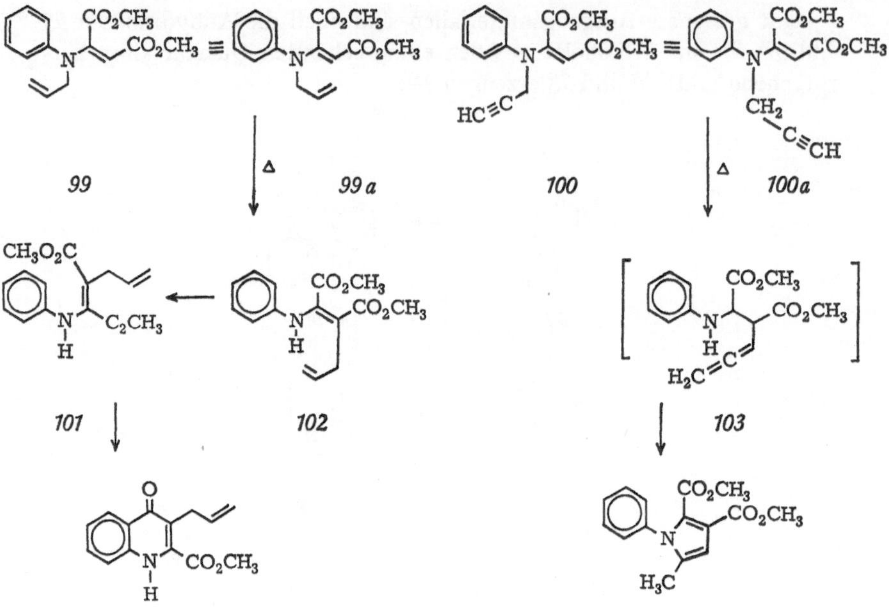

99   99 a   100   100 a

101   102   103

104   105

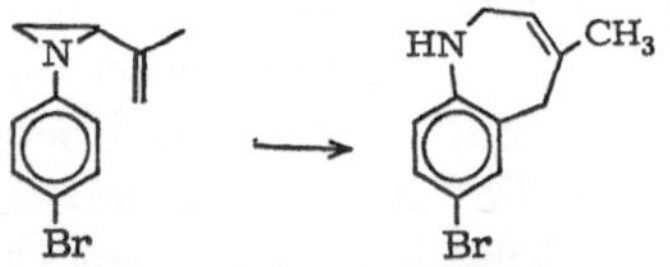

106 → 107

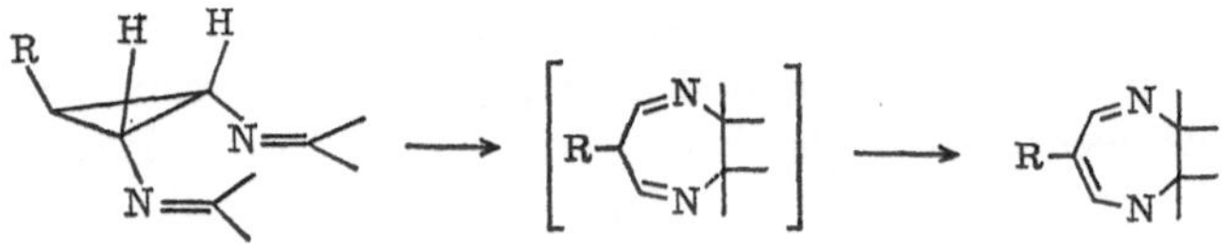

108 → 109 → 110

Diese Reaktion, die nicht die Primärprodukte *109*, sondern die konjugierten Systeme *110* liefert, sei an dieser Stelle erwähnt, obwohl sie nicht direkt als Claisen-Analogon, sondern vielmehr als eine Aza-Cope-Reaktion anzusehen ist. In diesem Zusammenhang müssen auch Valenzisomerisierungen vom Typ *111* → *112* gesehen werden, die von mehreren Arbeitskreisen untersucht wurden [56-59].

111   *112*

Wie G. Maier zeigte, liegt die Verbindung *113* nach Aussage des NMR-Spektrums zwar völlig in der Norcaradien-Struktur vor, da man aber beim Versuch der stereospezifischen Synthese nicht *114* sondern unabhängig vom Ausgangsmaterial nur *113a* erhält, wird eine ringoffene Zwischenstufe vom Typ *112* zur Erklärung dieses Effektes herangezogen (s. a. [58]).

113 R = H   *114*
113a R = CH$_3$

III. Fischer-Analoga

Dieses letzte Beispiel leitet bereits über zum Typ der Fischer-analogen Hetero-Cope-Reaktion, die dadurch charakterisiert ist, daß eine Bindung zwischen zwei Heteroatomen im einfachsten Fall, also zwischen zwei Stickstoffatomen, aufbricht. Die neueren Ergebnisse zur Synthese von Indolverbindungen sind in der ausführlichen Übersicht von Robinson [12] zusammengefaßt. Hier wird nur auf neuere und dort nicht berücksichtigte Entwicklungen hingewiesen. So erhält man aus 1,2-Dialkyl-phenyl-hydrazin *115* durch Umsetzung mit Ketonen der allgemeinen Struktur *116* beim Erhitzen auf 100—120 °C unter Abspaltung von Methylamin in Ausbeuten zwischen 25 und 80% die entsprechenden N-Alkyl-Indolderivate *117* [60].

115   *116*   *117*

Für präparative Vorhaben sollte jedoch unter Protonenkatalyse gearbeitet werden, weil sowohl eine geringere Reaktionszeit als auch eine durchweg höhere Ausbeute für diese Bedingungen angegeben werden.

Arbeitet man mit symmetrischem 1,2-Dialkyl-hydrazin, also ohne den Phenylrest, so kann dieses Verfahren zur Synthese von Pyrrolen Verwendung finden [61,62]. Die rein thermische Reaktion ist auch hier kein brauchbares Verfahren, unter schwacher Säurekatalyse bildet sich jedoch aus Cyclohexanon *118* und bei sukzessiver Einwirkung von Dimedon und Cyclohexanon *119*.

Die Reaktion versagt ihren Dienst beim Acetophenon, in mäßiger Ausbeute wird *120* isoliert.

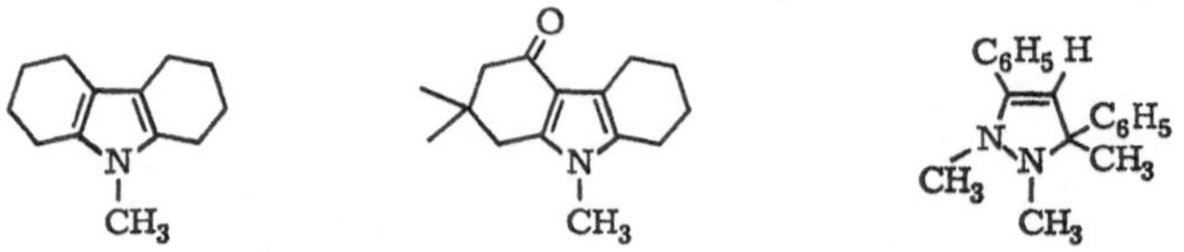

118   *119*   *120*

Das interessante und unerwartete Pyrrolderivat *124* gewinnt man bei der Umsetzung des 1,4-Diketons *121* mit Dimethylhydrazin. Der sigmatrope Primärprozeß des Kondensationsproduktes *122* liefert *123*,

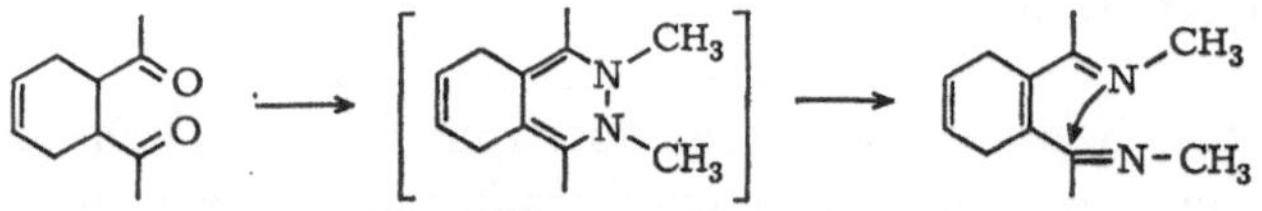

121   *122*   *123*

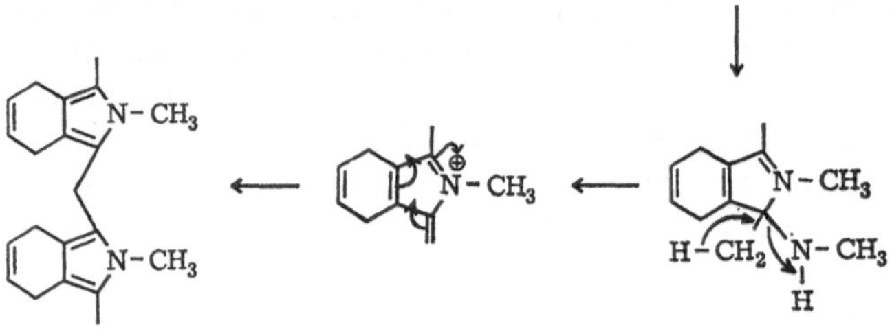

124   *125*

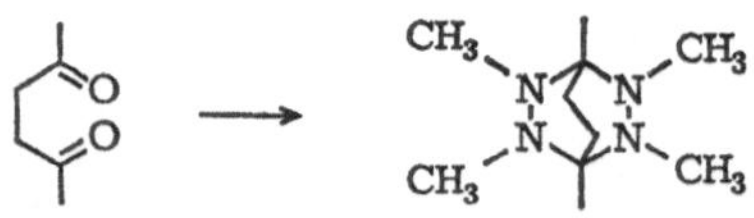

126   *127*

das unter Abspaltung von Methylamin ein zur Weiterkondensation wohl vorbereitetes Zwischenprodukt *125* erzeugt. Unter Aufnahme von 2 Mol Dimethylhydrazin und Bildung von *127* verläuft die Reaktion beim Hexandion-2,5 *(126)* [63].

Prozesse dieser Art, bei denen die Spaltung einer N-N oder einer N-O-Bindung die Umwandlung auslöst, sollten wegen der geringeren Bindungsenergie (38 Kcal/Mol für N-N und 48 Kcal/Mol für N-O gegenüber 83 Kcal/Mol für C-C) kleinere Aktivierungsenergien aufweisen. Neuere Resultate aus der Chemie der Hydroxylaminderivate bestätigen diese Erwartung.

O. House und F. A. Richey [64] fanden für das Enamin-acetat *129*, das sich aus dem durch Methylierung des Oximacetats darstellbaren Quartärsalz *128* mit Triäthylamin in wasserfreier Phase erzeugen läßt, bei 5—10 °C eine Halbwertszeit von weniger als 30 sec.

128　　　　　　　*129*　　　　　　　*130*

Die Enamin-Acetate lassen sich auch direkt aus N-Methylhydroxylamin-O-acetat und dem entsprechenden Keton gewinnen und zur Umlagerung bringen. Zur Selektivität dieser Reaktion ist hervorzuheben, daß das Acetat *131* unabhängig von der Oximkonfiguration nach Umlagerung und Hydrolyse stets zum gleichen Gemisch der stereoisomeren, sekundären Acetoxy-Ketone *132* und *133* führt.

132　　　　　　*133*　　　　　　*131*　　　　　　*134*

Die Eintopftechnik mit dem O-Acetat des N-Methylhydroxylamins jedoch liefert als Hauptprodukt das tertiäre Acetat *134*. House zieht den sicher berechtigten Schluß, daß offenbar die Konfiguration des Oxims bedeutungslos für die Richtung der Umlagerung ist, die vielmehr ausschließlich durch die Position der Enamin-Doppelbindung bestimmt wird.

Im Licht dieser Umlagerung muß wohl auch ein unerwartetes Resultat der Beckmann-Reaktion des Oxims *135* gesehen werden [65]. Unter den klassischen Beckmann-Bedingungen erhält man das Oxazol-Derivat *137*, und eine Deutung über *136* und *138* scheint in der Tat einleuchtend.

Langsam bei Raumtemperatur und rasch beim Erhitzen in Benzol verläuft auch die Umlagerung der aus dem Isoxazol-derivat *139* hervorgehenden Zwischenstufe *140*, wobei hier zusätzlich die Aktivierungsschwelle der Ringöffnung zu überwinden ist [66]. Isoliert wird das Enaminketon *141*, das durch seine besonders leicht eintretende Cyclisierung zum Pyrrol-tetracarbonester *142* charakterisiert ist.

Es sei jedoch ausdrücklich darauf hingewiesen, daß dies nicht der einzige Umlagerungsweg für die Pyrrolbildung aus Isoxazolderivaten ist. Unter bestimmten konstitutionellen Voraussetzungen kann der über

Aziridine verlaufende Weg als ausschließliche Alternative angetroffen werden [67,68]. So liefert zwar *143* noch teils über den Cope-Prozeß *144* die cyclische Form *(145)* des entsprechenden Enaminaldehyds, die kristallin isoliert und mit Säure in den Pyrrolester *146* überführt wurde,

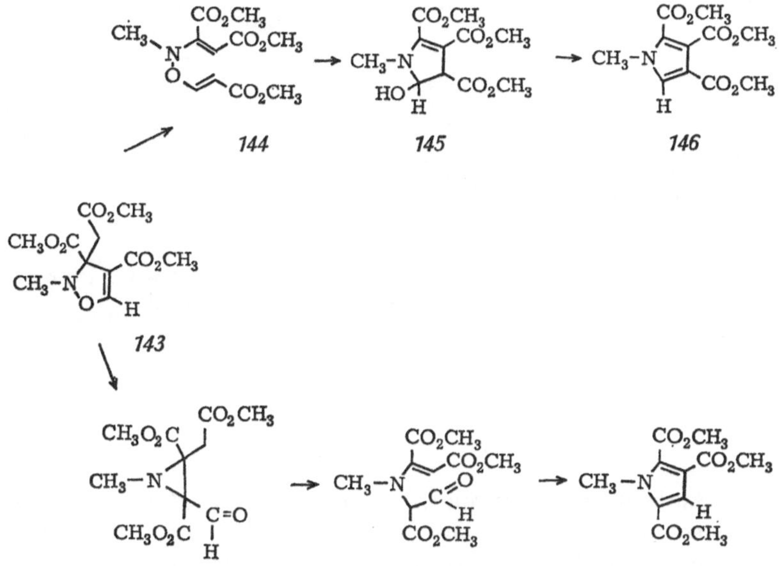

144   *145*   *146*

143

147   *148*   *149*

daneben bildet sich jedoch in etwa gleicher Menge der isomere Triester *149*, für dessen Bildung der Aziridin-Weg über *147* und *148* nahegelegt wird. Ausschließlich diesen Weg schlägt die Umlagerung des Diesters *150* ein. Als einziges Pyrrolderivat wird der Diester *153* isoliert. *151* und *152* bieten sich als Zwischenstufen an.

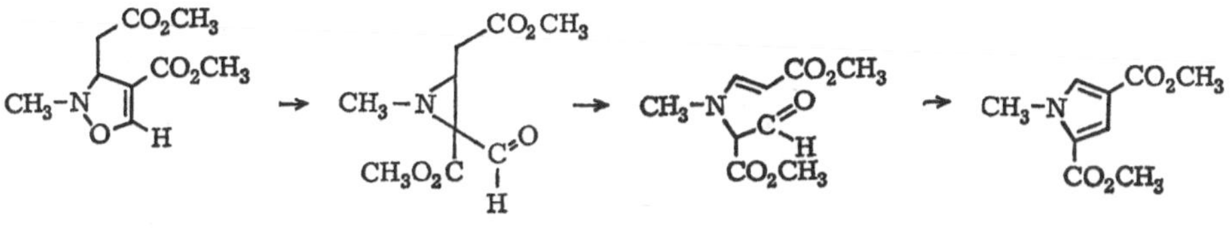

150   *151*   *152*   *153*

Das aus Phenylhydroxylamin und Acetylendicarbonester in hoher Ausbeute anfallende Isoxazol-Derivat *154* geht nach Hydroxylaminabspaltung und Ringöffnung in die Produkte *156* und *157* über [66]. Als Zwischenstufe drängt sich hier *155* auf, eine Struktur, der abweichend von den N-Alkylderivaten 2 Reaktionswege offenstehen.

Das Ausbeuteverhältnis von *156* und *157* (~ 1:4) zeigt, daß im Gegensatz zu den auf Seiten 87/88 angeführten Beispielen hier der Aromat in überwiegendem Maße am Umlagerungsprozeß beteiligt wird.

Große Ähnlichkeit mit dieser letzten Reaktion hat die zur Benzofuran-Bildung verwendbare Umlagerung vom Typ *158*, bei der über die Enamine *159* die Furane *160* gewonnen werden [70-73], sowie die kürzlich von Sheradsky mitgeteilte Synthese von Pyrolen *(163)* aus dem Oxim-

Derivat *(161)*. Hier wird eine thermische Enaminbildung angenommen, die dann das umlagerungsfähige System *162* bereitstellt [74].

Ebenfalls in diese Gruppe der Fischer-analogen Cope-Systeme mag schließlich noch ein mechanistisch sehr interessantes, wenn auch zunächst etwas spekulativ anmutendes Beispiel gehören, das zur Deutung der thermischen Umlagerung des Diphenylketen-Cycloadduktes *164* vorgeschlagen wurde [75]. Acceptiert man die etwas exotische 8-Ring-Struktur *165*, so wäre die Bildung des tatsächlich isolierten Endproduktes *(167)* über *166* ein plausibler Schritt.

IV. 1,3-ständige Heteroatome

Zwei Beispiele mit 1,3-ständigen Heteroatomen im 6-Zentren-System wurden bereits auf Seite 81 an Pyridin- und Chinolinderivaten angetroffen, und es nimmt nicht wunder, daß speziell die Reihe der Heterocyclen hier ein weites Feld von Möglichkeiten bietet. Ähnliche Resultate werden z. B. auch aus der Pyrimidinreihe mitgeteilt [76].

5-Ring-Heterocyclen sind ebenfalls ergiebige Substrate, wie ein kurzer Blick auf die Serie von Produkten aus Tetrazol-, Benzimidazol-, Benzoxazol- und Benzthiazolderivaten demonstriert [77,78].

Zur doppelten N-Alkylierung führt der sigmatrope Prozeß bei Purin-Äthern [79].

Carboxylderivate können natürlich auch die 1,3-ständigen Heteroatome bereitstellen, wie das Beispiel des Benzoesäureesters *177* lehrt. Bei 530 °C erhält man das Gemisch der cis, trans isomeren Nitrile *178* [80].

Dieser Reaktionstyp ist von amerikanischen Autoren detailliert untersucht worden [81-83], und anhand des O_{18}-markierten Crotylesters *179* ließ sich bei 268 °C in der Gasphase der cyclische Prozeß als Hauptweg überzeugend demonstrieren.

179   *180*

Auch die thermische Umlagerung des Chinol-Acetats *181*, die in Gegenwart von Pyridin neben Ausgangsmaterial 60% des Acetats *182* liefert, fällt in diese Gruppe, und erwartungsgemäß geht die nicht substituierte Verbindung in ein Phenol-Derivat über [84].

181   *182*

V. Ionische Übergangszustände

Auch bei geladenen Übergangszuständen bleibt der sigmatrope Prozeß symmetrie-erlaubt, und besonders bei Stickstoff-Verbindungen ist inzwischen eine Reihe umlagerungsfähiger Systeme untersucht worden. So konnte für den von Grob u. Mitarb. detailliert untersuchten Fragmentierungsprozeß *183* gezeigt werden, daß die tatsächliche Endstufe *185* aus dem primären Fragmentierungsprodukt *184* über eine rasche sigmatrope Folgereaktion gebildet wird [85,86].

183   *184*   *185*

Auch hier kann die Dreifachbindung in das Umlagerungssystem einbezogen werden, wie die von französischen Autoren untersuchte Bildung des Aldehyds *188* aus dem Morpholinderivat *186* lehrt [87].

Ein innermolekularer, cyclischer Prozeß konnte auch für die mit der Knabe-Umlagerung [88-90] eng verwandte Wanderung von Allyl- oder Propargylresten in Verbindungen vom Typ *189* und *192* gesichert werden [91,92], und es ließ sich auch das resultierende Imoniumsalz durch Einfang von Cyanid-Ionen stabilisieren [93].

Ebenfalls sehr rasch und mit praktisch quantitativer Ausbeute verläuft bereits bei Raumtemperatur eine sehr ähnliche Wanderung ungesättigter Reste bei den aus sekundären Aminen und Formaldehyd im sauren Medium spontan sich bildenden Imoniumsalzen *195* bzw. *198* [94].

Boranat-Reduktion überführt die Umlagerungsprodukte in die stabilen tertiären Amine *197* und *200*. Zur Erzeugung des Imoniumsalzes können auch andere aliphatische Aldehyde verwendet werden [95].

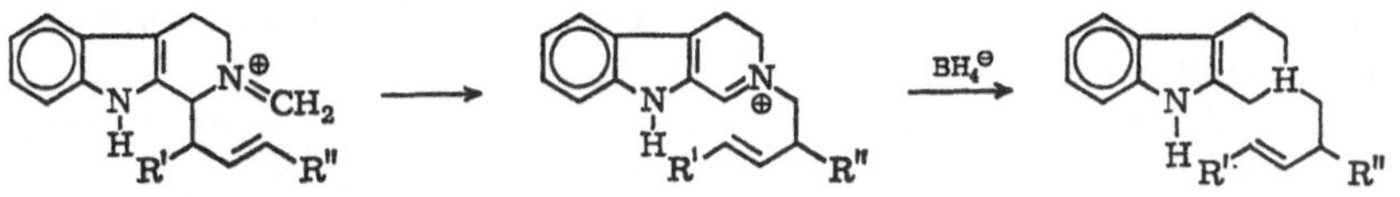

195   *196*   *197*

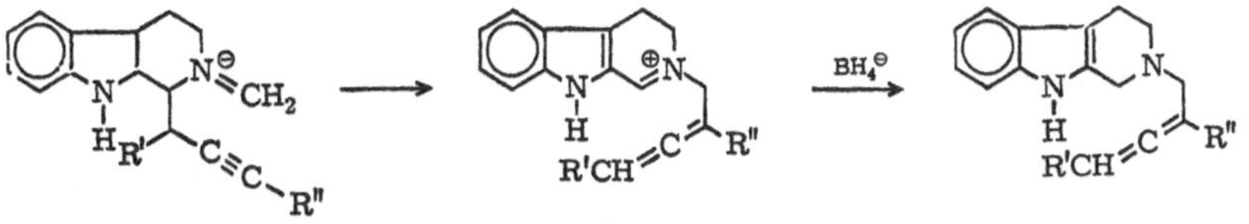

198   *199*   *200*

Während die Verbindungen *196* und *199* unter den Bildungsbedingungen stabil sind, durchlaufen die substituierten Produkte vom Typ *202* in Gegenwart von Donatoren wie Wasser oder Methanol eine stereoselektive Cyclisierung zu den Indolo-Chinolizidinen *203* [96].

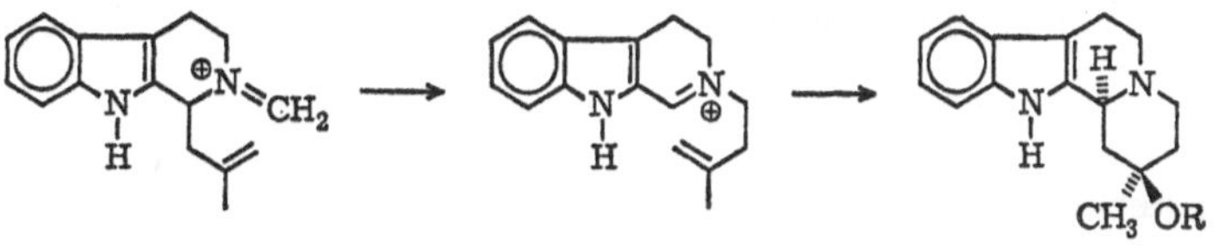

201   *202*   *203*

Zum Abschluß dieser Serie sei auf die mechanistisch interessante Umsetzung des Pyrrolidin-Enamins *204* mit Crotonsäurechlorid hingewiesen. Die Bildung des bei dieser Reaktion nach Hydrolyse isolierten Aldehyds *209* wird über eine Cope-Reaktion des Primäradduktes *205*

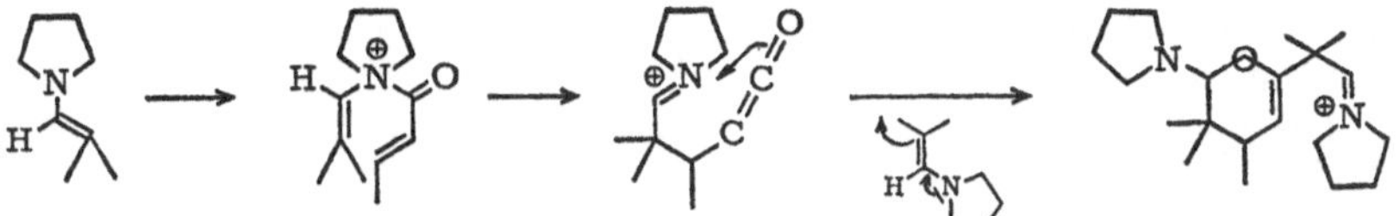

204   *205*   *206*   *207*

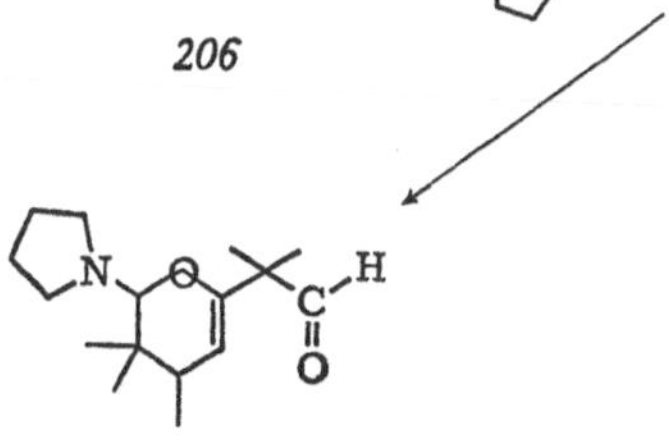

208

und einen Abfang des sich bildenden Ketens *206* durch ein zweites Mol des Enamins verstanden [97].

Die Bildung der C-C-Bindung durch nucleophilen Angriff der Enamin-doppelbindung auf das β-C-Atom des ungesättigten Säurechlorids scheint jedoch nicht mit letzter Sicherheit ausgeschlossen.

VI. Ausblick

Es ist der Wunsch, daß diese Auswahl von Hetero-Cope-Reaktionen, die der neueren Literatur entnommen wurde und keinen Anspruch auf Voll-ständigkeit erhebt[a], dazu anregt, sich der großen Variationsfähigkeit des unter *1* angegebenen Reaktionsschemas bewußt zu sein. Die zitierten Beispiele legen zum größten Teil Erweiterungen und Übertragungen auf andere Moleküle und Verbindungsklassen nahe, und es ist zu erwarten, daß diese Lücken in den kommenden Jahren geschlossen werden. Dabei werden sicher neben vielen mechanistisch interessanten Umwandlungen auch eine große Zahl präparativ nützlicher und ergiebiger Reaktionen als Lohn für Mühe und Fleiß der Experimentatoren aufgefunden werden.

VII. Literatur

[1] Cope, A. C., Hardy, E. M.: J. Am. Chem. Soc. *62*, 441 (1940).

[2] Doering, W. v. E., Roth, W. R.: Tetrahedron *18*, 67 (1962)

[3] Hammond, G. H., de Boer, C. D.: J. Am. Chem. Soc. *86*, 899 (1964).

[4] Trecker, D. J., Henry, J. P.: J. Am. Chem. Soc. *86*, 902 (1964).

[5] Woodward, R. B., Hoffmann, R.: J. Am. Chem. Soc. *87*, 395 (1965).

[6] — — J. Am. Chem. Soc. *87*, 2046 (1965).

[7] — — J. Am. Chem. Soc. *87*, 2511 (1965).

[8] — Aromaticity Special Publication. J. Chem. Soc. 217 (1967).

[9] — Hoffmann, R.: Angew. Chem. *81*, 797 (1969).

[10] Zusammenfassung thermischer Prozesse in der Kohlenstoff-Serie s. Frey, H. M., Walsh, R.: Chem. Rev. *1969*, 103.

[11] Jefferson, A., Scheinmann, F.: Quart. Rev. (London) *1968*, 391.

[12] Robinson, B.: Chem. Rev. *1969*, 227.

[13] le Noble, W. J., Gabrielsen, B.: Chem. Ind. (London) *1969*, 378.

[14] Frater, G. Y., Habich, A., Hansen, H. J., Schmid, H.: Helv. Chim. Acta *1969*, 335 (weitere Zitate s. dort).

[14a] Hansen, H. J., Schmid, H.: Chimia *24*, 89 (1970).

[15] Thomas, A. F.: Chem. Commun. *1967*, 946.

[16] — Ozainne, M.: Chem. Commun. *1970*, 220.

---

[a] Die Auswahl ist durch die speziellen Interessengebiete des Autors geprägt. Auto-ren, deren Beiträge zu diesem Gebiet nicht oder nicht in vollem Maße berück-sichtigt wurden, werden an dieser Stelle um freundliche Nachsicht gebeten.

17) Mutterer, F., Morgen, J. M., Biedermann, J. M., Fleury, J. P., Weiß, F.: Tetrahedron *26*, 477 (1970).
18) Johnson, W. S., Werthemann, L., Bartlett, W. R., Brocksom, T. Z., Tsung Tee Li, Faulkner, D. Z., Petersen, M. R.: J. Am. Chem. Soc. *92*, 741 (1970).
19) Hoffmann, W., Pasedach, H., Pommer, H.: Liebigs Ann. Chem. *729*, 52 (1969).
20) Faulkner, D. J., Petersen, M. R.: Tetrahedron Letters *1969*, 3243.
21) Wakabayashi, N., Waters, R. M., Church, J. P.: Tetrahedron Letters *1969*, 3253.
22) Dinan, F. J., Tieckelmann, H.: J. Org. Chem. *29*, 892 (1964).
23) Moffet, R. B.: J. Org. Chem. *28*, 2885 (1963).
24) Makisumi, Y.: Tetrahedron Letters *1964*, 2833.
25) Claisen, L.: Chem. Ber. *45*, 3157 (1912).
26) Win, H., Tieckelmann, H.: J. Org. Chem. *32*, 59 (1967).
27) McLamore, W. M., Gelbum, E., Rauley, A.: J. Am. Chem. Soc. *78*, 2816 (1956).
28) Hurd, C. D., Trofimenko, S.: J. Org. Chem. *23*, 1276 (1958).
29) Dashunin, V. M., Tovbina, M. V.: J. Gen. Chem. USSR *34*, 1443 (1964).
30) Makisumi, Y.: J. Synth. Org. Chem. Japan *27* (7) 593 (1969).
31) — Murabayashi, A.: Tetrahedron Letters 1971 (1969).
32) — Sasatani, A.: Tetrahedron Letters 1975 (1969).
33) Kwart, H., Evans, E.: J. Org. Chem. *31*, 413 (1966).
34) — Cohen, M. H.: J. Org. Chem. *32*, 3135 (1967).
35) — Schwartz, J. L.: Chem. Commun. *1969*, 44.
36) Makisumi, Y., Marabashi, A.: Tetrahedron Letters *1969*, 2449.
37) — — Tetrahedron Letters *1969*, 2453.
38) Brandsma, L., Bos, H. J. T.: Recueil *88*, 732 (1969).
39) Bycroft, B. W., Landon, W.: Chem. Commun. 168 (1970).
40) Baldwin, J. E., Brown, J. E., Cordell, R. W.: Chem. Commun. 31 (1970).
41) — Urban, F. J.: Chem. Commun. 165 (1970).
42) — Debernardis, J., Patrick, J. E.: Tetrahedron Letters 353 (1970).
43) Schuijl, P. J. W., Bos, H. J. T., Brandsma, L.: Recueil *88*, 597 (1969).
44) — Brandsma, L.: Recueil *87*, 929 (1968).
45) Schuler, F. W., Murphy, G. W.: J. Am. Chem. Soc. *75*, 3155 (1950).
46) Kincaid, J. F., Tarbell, D. S.: J. Am. Chem. Soc. *61*, 3085 (1939).
47) Wijert H. E., v. Ginkel, C. H. D., Schuijl, P. J. W., Brandsma, L.: Recueil *87*, 1136 (1968).
48) Hill, R. K., Gilman,N. W.: Tetrahedron Letters 1421 (1967)
49) Makisumi, Y.: Tetrahedron Letters 6413 (1966).
50) — Tetrahedron Letters 543 (1969).
51) Hill, R. K., Newkome, G. R.: Tetrahedron Letters 5059 (1968).
52) Schmidt, G.: Dissertation, Technische Universität Berlin 1970.
53) Scheiner, P.: J. Org. Chem. *32*, 2628 (1967).
54) Staab, H. A., Vögtle, F.: Tetrahedron Letters 51 (1965).
55) — — Chem. Ber. *98*, 2681, 2691, 2701 (1965).
56) Maier, G.: Chem. Ber. *95*, 611 (1962); *98*, 2438, 2446 (1965).
57) Amiet, R. G., Johns, R. B., Markham, K. R.: Chem. Commun. 128 (1965).
58) Sauer, J., Heinrichs, G.: Tetrahedron Letters 4979 (1966).
59) Turner, A. B., Heine, H. W., Irving, J., Bush, J. B.: J. Am. Chem. Soc. *87*, 1050 (1965).
60) Schieß, P., Grieder, A.: Tetrahedron Letters 2097 (1969).
61) Sucrow, W.: Chimia *23*, 36 (1969).
62) Jaquier, R., Chapelle, J. P., Elguero, J., Tarrago, G.: Chem. Commun. 752 (1969).
63) Baldwin, J. E., Basson, H. H.: Chem. Comm. 795 (1969).
64) House, O., Richey, F. A.: J. Org. Chem. *34*, 1430 (1969).

65) Galantey, E., Hoffman, C., Paolella, N.: Chem. Commun. 274 (1970).

66) Winterfeldt, E., Krohn, W., Stracke, H. U.: Chem. Ber. *102*, 2346 (1969).

67) Baldwin, J. E., Pudussery, R. G., Qureshi, A. K., Sklarz, B.: J. Am. Chem. Soc. *90*, 5325 (1968).

68) Adachi, I., Harada, K., Kano, H.: Tetrahedron Letters 4875 (1969).

69) Schmidt, G.: Dissertation, Technische Universität Berlin 1970.

70) Sheradsky, T.: Tetrahedron Letters 5225 (1966).

71) Mooradian, A.: Tetrahedron Letters 407 (1967).

72) Kaminsky, D., Shavel, J., Meltzer, R. I.: Tetrahedron Letters 859 (1967).

73) Mooradian, A., Dupont, P. E.: Tetrahedron Letters 2867 (1967).

74) Sheradsky, T.: Tetrahedron Letters 25 (1970).

75) Bird, C. W.: Chem. Commun. 1486 (1969).

76) Minnemeyer, H. J., Clarke, P. B., Tieckelmann, H.: J. Org. Chem. *31*, 406 (1966).

77) Elwood, J. K., Gates, J. W.: J. Org. Chem. *32*, 2956 (1967).

78) Takahashi, S., Kano, H.: Chem. Pharm. Bull. (Tokyo) *12*, 282 (1964).

79) Bergmann, E., Heimhold, H.: J. Chem. Soc. *1935*, 1365.

80) Holm, T.: Acta Chem. Scand. *1965*, 242.

81) Snyder, H. R., Stewart, J. M., Myers, R. L. J.: J. Am. Chem. Soc. *71*, 1055 (1949).

82) Lewis, E. S., Hill, J. T., Newsman, E. R.: J. Am. Chem. Soc. *90*, 662 (1968).

83) — — J. Am. Chem. Soc. *91*, 7458 (1969).

84) Barton, D. H. R., Magnus, P. D., Pearson, M. J.: Chem. Commun. 550 (1969).

85) Geisel, M., Grob, C. A., Wohl, R. A.: Helv. Chim. Acta 2206 (1969).

86) Marshall, J. A., Babler, J. H.: J. Org. Chem. *34*, 4186 (1969).

87) Cresson, P., Corbier J.: Compt. Rend. C *268*, 1614 (1969).

88) Knabe, J., Rupenthal, N.: Arch. Pharm. *299*, 129 (1966).

89) — — Naturwissenschaften *51*, 482 (1964).

90) — Detering, K.: Chem. Ber. *99*, 2873 (1966).

91) — Tetrahedron Letters *433*, 2107 (1969).

92) Brown, D. W., Dyke, S. F., Kinsman, R. G., Sainsbury, M.: Tetrahedron Letters 1731 (1969).

93) Sainsbury, M., Brown, D. W., Dyke, S. F., Kinsman, R. G., Moon, B. J.: Tetrahedron 6695 1968).

94) Winterfeldt, E., Franzischka, W.: Chem. Ber. *100*, 3801 (1967).

95) — — Chem. Ber. *101*, 2938 (1968).

96) Rischke, K.: Dissertation, Technische Universität Berlin 1970.

97) Hickmott, P. W., Hopkins, B. J.: J. Chem. Soc. C *1968*, 2918.

Received April 23, 1970

SPRINGER-VERLAG
BERLIN·HEIDELBERG·NEW YORK

Dritte, völlig
neubearbeitete
Auflage

D'Ans/Lax
Taschenbuch für
Chemiker und Physiker

Herausgegeben von Professor Dr. **Klaus Schäfer,**
Physikalisch-Chemisches Institut der Universität
Heidelberg, und Dr. **Claudia Synowietz,** Heidelberg

Band III

Eigenschaften von
Atomen und Molekeln

Mit 112 Abbildungen
VIII, 670 Seiten. 1970
Gebunden DM 48,–
US $ 13.20

Inhaltsübersicht: Grundkonstanten des Atomismus. — Elementarteilchen. — Isotopengewichte, Massedefekte, Packungsteile und Kerneigenschaften. — Natürliche und künstliche Radioaktivität. — Kernmagnetische Resonanz und chemische Verschiebung. — Periodensysteme und Grundterme. — Spektren. — Ionisierungsspannungen und Elektronenaffinitäten. — Ionenradien. — Atomfaktoren und Wirkungsquerschnitte in Gasen. — Kernabstände und Valenzwinkel in Molekülen. — Polarisierbarkeiten. — Dipolmomente. — Konstanten. — Mechanik und Quantenstatistik. — Kinetische Gastheorie. — Kristallographische Gitterstrukturen. — Verzeichnis der Spektrallinien für analytische Zwecke. — Sachverzeichnis.

■ **Bitte Prospekt
anfordern!**

Früher erschienen

Band II

Organische Verbindungen
VIII, 1177 Seiten. 1964
Gebunden DM 48,—; US $ 13.20

Band I

Makroskopische physikalisch-
chemische Eigenschaften
Mit 350 Abbildungen. XVI, 1522 Seiten. 1967
Gebunden DM 68,—; US $ 18.70